WHAT IS MUSIC PRODUCTION?

WHAT IS MUSIC PRODUCTION?

A producer's guide: the role, the people, the process

Russ Hepworth-Sawyer

Craig Golding

AMSTERDAM • BOSTON • HEIDELBERG • LONDON
NEW YORK • OXFORD • PARIS • SAN DIEGO
SAN FRANCISCO • SINGAPORE • SYDNEY • TOKYO

Focal Press is an imprint of Elsevier

Focal Press is an imprint of Elsevier
30 Corporate Drive, Suite 400, Burlington, MA 01803, USA
The Boulevard, Langford Lane, Kidlington, Oxford, OX5 1GB, UK

Notices
Knowledge and best practice in this field are constantly changing. As new research and experience broaden our understanding, changes in research methods, professional practices, or medical treatment may become necessary.

Practitioners and researchers must always rely on their own experience and knowledge in evaluating and using any information, methods, compounds, or experiments described herein. In using such information or methods they should be mindful of their own safety and the safety of others, including parties for whom they have a professional responsibility.

To the fullest extent of the law, neither the Publisher nor the authors, contributors, or editors, assume any liability for any injury and/or damage to persons or property as a matter of products liability, negligence or otherwise, or from any use or operation of any methods, products, instructions, or ideas contained in the material herein.

Library of Congress Cataloging-in-Publication Data
Hepworth-Sawyer, Russ.
A producer's guide : the role, the people, the process / Russ Hepworth-Sawyer, Craig Golding.
    p. cm.
ISBN 978-0-240-81126-0
1. Sound recordings–Production and direction. I. Golding, Craig. II. Title.
ML3790.H493 2010
781.49–dc22

2010039555

British Library Cataloguing-in-Publication Data
A catalogue record for this book is available from the British Library.

ISBN: 978-0-240-81126-0

For information on all Focal Press publications
visit our website at www.elsevierdirect.com

10   11   12   13   5   4   3   2   1

Printed in the United States of America

Dedication

To Lucy, Ben, and Tom

Contents

Section A ● What Is Music Production?

Section B ● Being It

Section C ● Prepping It

Section D ● Doing It

Section E ● The Future

Section F ● Appendices

RUSS HEPWORTH-SAWYER

Russ Hepworth-Sawyer is a sound engineer and producer with many years' experience of all things audio and is a member of the Association of Professional Recording Services and the Audio Engineering Society; a Fellow of the Institute For Learning (U.K.); and a board member of the Music Producer's Guild. Through MOTTOsound (*www.mottosound.co.uk*), Russ works freelance in the industry as a mastering engineer, a producer, writer, and consultant. Russ currently lectures part-time for York St John University and Barnsley College Online and has taught extensively in higher education at British institutions, including Leeds College of Music, London College of Music, and Rose Bruford College. He currently writes for *Pro Sound News Europe*, has contributed to *Sound On Sound* magazine, and has written many titles for Focal Press.

CRAIG GOLDING

Craig Golding is currently Course Leader for the BA (Hons) Music Production degree program at Leeds College of Music in the U.K. Craig lectures in production, sound recording techniques, studio production, and song production and leads a team that is responsible for the delivery and development of one of the most popular music production courses in the U.K. Craig also has an active freelance career in sound engineering and production with over a decade's experience working in the industry. Under the banner of CMG Sound Craig has worked as an FOH engineer at many large venues in the U.K. including The Sage Gateshead; The Bridgewater Hall, Manchester; and Queen Elizabeth Hall, London. Craig continues to work with a variety of artists and ensembles both in live and studio contexts.

Acknowledgments

The authors would like to jointly acknowledge…

Danny Cope, Stefan Gordon, and most importantly Ben Burrows for taking the time to check our writing, help us edit, and bring this book to fruition. Thanks, chaps. Thanks must go to Helen Golding for the cover artwork—thanks, Helen. We'd also like to mention the persistence of Catharine Steers and her team at Focal Press for their support and patience through the writing process. Additional thanks to Sam Molineaux-Graham for copyediting our text and helping us with the US relevance.

However, more importantly, we simply could not have written this book without the help, assistance, professionalism, and passionately driven inspiration of the producers and music professionals we've had the pleasure to discuss this topic with recently and over many years. Thank you.

In addition we'd wholeheartedly like to thank the following music professionals for their time, support, and contribution to this book. You are all an inspiration: Dave Aston; Tom Bailey; Andy Barlow; Haydn Bendall; Leonard Bendell; Danny Cope; Tommy D; Joe D'Ambrosio; DJ Jon DaSilva; Will Francis; Mick Glossop; Stefan Gordon; Phil Harding (for providing a fantastic foreword—thanks, Phil); Joel Harrison; Rupert Hine; Andrew Hunt; Paul K. Joyce; Bob Katz; Richard Lightman; Barkley McKay; Richard Mollet; Brian Morrell; Robert Orton; Tony Platt; Tim "Spag" Speight; Ray Staff; Malcolm Toft; and the many names we've erroneously forgotten from this list.

This book was written using Google Docs.

Russ would like to thank…

Thanks must first go to my wife Jackie and son Tom for allowing me to embark on yet another book—thank you for your love and encouragement! Thank you, Mum too for getting me that first guitar all those years ago and thanks to my brothers, especially Ray and Peter, for coming to all those gigs when you must have hated the music! Thanks too must go to my parents-in-law Ann and John for all their help and support. Utmost respect is due to Craig Golding, who I am sure at times wished he'd never entered the book-writing game but without whom this book would never have happened. Thanks to Max Wilson for his friendship, honesty, and sincere support over all these years… thanks, boss! Continual thanks owed to: Iain Hodge and Peter Cook at London College of Music; Danny Cope, Barkley McKay, and Brian Morrell and former colleagues at Leeds College of Music; and Ben Burrows and Rob Wilsmore and colleagues at York St John University. Thanks also to my friends in the industry: Dave Aston,

Tom Bailey, Paul Baily, Len Bendall, Mark Cousins, Will Francis, Rob Orton, and Ray Staff for their help and encouragement in moving the project along over many years. Also a quick shout-out to Ruth D'Silva; Olivia Flenley; Sue, Toto, and Billy Sawyer; Catherine "Parsonage" Tackley; and Tony Whyton. Thanks are also due to my colleagues at the Music Producers Guild for their individual support during the writing of this book. I'd also like to thank all the students I have taught over the many years—you have taught me more!

Craig would like to thank…

My first thank you has to go to my wife Joanne, my son Ben, and my daughter Lucy. "The book" has had a bigger impact on you than it ever will on anyone else—and you'll probably never read it! Thank you for your love and support and for giving me the time and space to write. Thanks to my parents, Mum, Dad, without you I couldn't have even read a book, never mind write one! My sister Helen, thanks for your talents and the use of your home as my second office! My sincere thanks to Russ Hepworth-Sawyer for persuading me that it would be great to write a book; thanks for trusting me. It's great to finally cross the finishing line together! My sincere thanks and appreciation to all the professionals who allowed me to interview them. Continual thanks and gratitude to all the music production team and colleagues at Leeds College of Music; to Paul Baily; Bhupinder Chaggar; Danny Cope; Alex Halliday; Ally Jowett; Barkley McKay; Brian Morrell; Neil Myers; Charlotte Orba; Dale Perkins; Jez Pritchatt; and Randall Whittaker. It is my privilege to work with such talented individuals. And finally, thanks to the students I have taught and worked with over the years—I hope I made a difference!

Foreword

It would be madness to suggest that a definitive book could be written about the music producer that encompassed all musical genres. Logic might suggest that a collection of books should be written, possibly one for each music genre since the job of a classical music producer is quite different to that of a pop music producer. Or perhaps a series of books ought to be written for each style of music *producer*?

Richard Burgess's *The Art of Record Production* and Howard Massey's *Behind the Glass* are long-standing texts in this area. Both books, in my view, capture the essence of the modern music producer in the rock and pop genres. What Russ and Craig have brought to the mix is a unique process-driven focus on the role, day-to-day responsibilities both in and out of the studio, and the future of the music producer.

I've known Russ and Craig for some time as I have regularly visited the institutions at which they teach, including Leeds College of Music, to deliver guest lectures to undergraduate and post-graduate students studying music production. In this book, just as in their lectures, they have set about the task of explaining what music production is in a very logical and relevant manner. They begin by quantifying the role of the music producer, moving on to the nuts and bolts of preparing for sessions, managing the sessions, contemplating deals, preparing for the final stages of mastering, and the importance of follow-up networking.

Music production has changed enormously over the last 30 or 40 years. Today's technology now allows anyone to make a high-quality recording at home and even release it to the public via the internet, all in a very short space of time. That said, the analog technologies and techniques of record making from the pre-digital era are still with us, and inform much of the processes used today. It is important for anyone wishing to enter the game in the 21st century to possess the contextual background of the industry as it once was as well as the current possibilities. Over and above this is the need to understand the skills, attributes, and the being of the music producer—all of which Russ and Craig achieve in this excellent and comprehensive "must-read" book.

Well done for taking on the task, chaps.

Phil Harding

Producer and Engineer, Suffolk, U.K.

Introduction

Music production *per se* is a strange entity and an odd thing to do. It is certainly a challenge to quantify and very difficult to measure, analyze, and then simulate. Despite this, we perceived a need for a guide that provided an insight into the world of record production, as requested time and again by our students and various other people.

This book tries to offer not only this insight but also some solid guidance. It does not try to over-analyze seminal albums and the actions of the producers, engineers, or musicians involved. Neither does it seek to over-glorify these oft unsung heroes; however, we must admit we might admire a few of them along the way. What we *are* interested in is providing an overview of the process of making music work and the day-to-day activity of the producer and his or her colleagues. The nuts and bolts, if you like, and a detailed look at this fascinating and rewarding profession.

Using real-world examples and a few interview excerpts, we delve into approaches to the production process and the skills required to perform the various tasks. More than anything, this book is a production "guide." There is neither an overbearing amount of interview material nor overblown step-by-steps on producing. The range of professionals whose words and experiences are woven into this book can be seen in the acknowledgments section before this introduction; we offer some of their shared themes and also their diverging viewpoints. As you'll read, not all agreed with the majority view, and by presenting their opinions we discover that production can be many things to many people.

Getting to the status of a producer can be a long and arduous endeavor. Once, and only if, success arrives, the producer has further work to do to maintain momentum and carry this forward. Like many other careers, the road is rocky and contains pitfalls. We touch on some of the more practical solutions to staying on top of the game, gleaned from friends and colleagues who have graciously given their time to be interviewed for this book. We wrote this book because of them, and because of all the production professionals before them and after (which includes you!).

Whatever your slant on music production and its variety of guises, many of the processes involved are about guiding the events, performances, people, and magic that result in a musical story.

This book is a guide to that guidance.

Russ Hepworth-Sawyer and Craig Golding

Northern England, U.K.

What is Music Production. DOI: 10.1016/B978-0-240-81126-0.00023-8

SECTION A
What Is Music Production?

CHAPTER A-1
Quantifying It

As we inducted new students to their degree in music production, we would find year on year that something odd happened. These induction sessions often began by posing one simple question: "What is a music producer?"—the same single question often asked at their interviews. The answers from both the interviews and the hour-long induction discussion ranged from bizarre to more obvious. One such odd one was "miracle worker," another was "the fixer" whilst others revolved around the more standard "make things sound good" or "make sure the band don't get too drunk" among many other attributes.

After many years of few concrete answers we set about asking whether we could externalize what we thought music production was, the way the degree saw it, and the way the institution perceived it. We realized no one really answered it. That is, answered it the way we might answer it.

This book is our response to that. It seeks not to express how to record a guitar, what chords to play in answer to the last chorus, or whether to change the song's tempo from 120 to 140 bpm. *What Is Music Production* does represent the often neglected additional skills and actions undertaken in the course of a producer's day.

WHAT IS MUSIC PRODUCTION?

Music production over the years has become a term that refers to a multitude of different skills seemingly thrown together to create a role. However, this role has developed far beyond its understood meaning of just some 15 years ago. Therefore we can simply deduce that the role is ever evolving and emerging. In order to get there, we need to take a quick look at some of what has come before.

Looking back to before the mid-1960s, the role of producer consisted of being a fixer (booking artists, musicians, and studios), A&R (Artist and Repertoire), plus the ultimate manager of time and resources. We acknowledge that Sir George Martin is perhaps an icon for change to this model, and many may still naturally cite Sir George Martin as the best example of a music producer. Martin was

What is Music Production. DOI: 10.1016/B978-0-240-81126-0.00001-9

already very successful (leading the Parlophone label with a number of classical recordings and comedy artists such as Peter Sellers, Peter Cook, and Spike Milligan) before taking on the Beatles, who had been turned down by other labels.

His work was to develop and nurture the creative and musical forces within the Beatles, to such an extent that many considered him to be the fifth member. He arranged and wrote string parts and was involved in some of the technical processes behind the sounds that were created. Martin focused his skills on producing the Beatles for many years and created a new perspective on music production, which would have a long-lasting effect. Many books have followed this exceptional contribution, and we will not dwell on Martin's career specifically here, but what effect he has had on the producer's role.

With no disrespect to Sir George, nor to the superb work he has achieved, it is no longer possible to identify him as the best example of music production as it stands today. It is perhaps controversial to think like this, but the role of the producer has come to mean so many different things today, something we hope this book explores.

In Sir George Martin's many influential years with the Beatles at Abbey Road Studios in London, he developed a working relationship and style that became arguably perceived as *modus operandi* for many years, and to some extent many rock and pop genres today still follow this lead—so much so that we refer to this style of work as that of the "traditional producer."

THE TRADITIONAL PRODUCER

We refer to the traditional producer as someone who has been allowed creative control of a recording process. For example, we would presume that a traditional producer working throughout the 1960s to the 1980s would have been used to working sometimes with a larger engineering team consisting of an engineer (if required or chosen) and perhaps a tape operator and/or assistant engineer. Their role would be to capture and encourage the performances of the artists using the options available to them. The producer would be the soundboard for the artist: someone to bounce ideas off and to receive an objective opinion from.

So what is music production then? Is it what existed before the '60s? Is it what George Martin and his contemporaries created? Is it the excesses traditionally attributed to the '70s and '80s rock culture? Or is it something completely different?

So what is music production?

"A lot of the concept of record production and what a producer is has really changed in the last five years... What makes you a record producer these days?"

British Producer, Tommy D, in interview.

We have concluded, through our research and discussions over the years, that *guidance* is a common denominator. Producers guide the process, guide the people, nurture the talent, and enhance the music. This book is thus a guide to that activity. A guide to guidance, if you like.

In the traditional sense, the producer's creative role is to develop the artist's music to a level at which it can be realized. This realization could take the form of being a commercial release, where the impetus is sales and exposure, or it could take the form of being artistically realized, in that the impetus is to achieve something unique or innovative whether it sells or not. Either way the producer's role is important and often misunderstood for those outside of the career. To suggest that the role is a solely creative force in the recording process would be incorrect. There are many other aspects less fun and influential that are just as important, which we'll cover as we go through the book.

"I'm a producer"

During the first session of an evening course we taught, we held a jovial introductory session where each student introduced who they were and why they were in the course. Students simply responded, "I'm a guitarist in a band and I wanted to learn how to get better recordings." There is nothing wrong in that statement, as that is partly why the course exists. Others said, "I'm a producer." In the past no exception would have been taken to that term, as we interpret two meanings of it. However, it does display a marked change in the way in which music creation is now perceived and carried out. One assumes a producer (akin to that of the traditional producer) would not require such a course. This helps to highlight the important fact that there are two meanings to this term.

Bedroom producers, as they have become dubbed by many, have indeed become not only hugely successful in their own right, but have also shown the world they are true producers in the historical sense of the word when they have come out of the bedroom. One such example, if the media hype is to be believed, is Chipmunk (a.k.a. Jamaal Fyffe) who at the age of 18 began an album with a friend in a bedroom, and with some slick assistance (perhaps from a seasoned producer and engineer) created the 2009 masterpiece *I Am Chipmunk,* upon which were the hit UK singles "Oopsy Daisy" and "Diamond Rings."

Another famous early example was White Town (a.k.a. Jyoti Mishra) who created "Your Woman" in his Derby home in the UK, which made it to No. 1 in 1997. However, in fairness, there have been many albums recorded at homes prior to this (think Les Paul, for example), but this shone the light squarely at what could be done with very little money and a bedroom studio married with modest modern digital technology.

The seemingly derogatory term of bedroom producer has been born over the last decade or so and it seeks to describe the person who sits at home, plugging away at a sequencer (or as it has developed to these days, a digital audio workstation or DAW) "producing" music. Let's be clear—the word *produce* here can be used in its meaning of making something sonically as opposed to guiding its creativity.

David Gray began his career based on several albums worth of recordings in a bedroom studio. One track, with some of that "slick" assistance and vocal rerecordings, created what became "Babylon" on *White Ladder*"(1999). "Babylon" went on to become the bestselling single of all time in Ireland and led to a Best New Artist Grammy Nomination in 2002.

The term "producer" took on a new meaning for many with the huge rise of dance culture during the '90s. It was cool to have decks and claim to be a producer in your spare time, with many actually making music their career. This was an explosion based out of new musical technology such as the widespread availability of sequencers, samplers, and synthesizers all communicating by MIDI. Naturally this book touches upon the art involved in this type of production, as it is music creation in its own right and should not be dismissed.

Whilst we discuss this type of producer from time to time, it is important to consider the context and therefore the role of the traditional producer is more our main focus, as well as where things have come from, and the art of production. The mystique and skills therein are explored, so that they can be utilized by all producers whether in the increasingly competent bedroom or in the elaborate studio complex.

The role of the producer

To try to define the role of the producer is to try to comprehend the sheer range of skills and attributes that come together. One could suggest it essentially encompasses a wide range of skills in one job title combined with some incredible pressures and vital talent. In some ways the pages of this book are a good example of this. Fingering through the contents pages will give a flavor of the wide range of activities and skills required. These range from musical skills, to audio engineering, consultant, counselor, financial manager, project manager, and so on. We spend some time in Section B looking into this and later in pertinent chapters along the way.

The creative input of the producer is unquestionable. The produced style of the music, artistically positioned within its genre of the time, could be the trait of that producer, something we refer to later in this book as the producer's *watermark*. This watermark is placed on the music either deliberately or subtly, giving it an identity bigger or more substantial than the component parts. From the producer's perspective, this is their trademark and could be considered their calling card. The saying is often recited, "you're only as good as your last record" and can be held true for so many producers of the day who then are relegated to the amusing, yet apt, "where are they now pile" (Spinal Tap, 1984).

Seriously though, we can all identify with an album from our youth or many years ago that still holds value in our hearts. This will perhaps be partly because of the association with our lives at the time, but much of the value may come from the quality of the music and its production at that time. Spotify, the free music streaming service, has been a revelation as it allows the listener to revisit music long since packed away with the age-old vinyl record collection in the loft. The dust has been cast aside and this material now breathes with new life. Yes, we dare admit much of it was pretty awful and it deserves the dust, but much can assist in identifying production gems, highlighting skills and processes which simply worked for that time, place, and genre.

Recognizing and analyzing the producer's watermark in more depth is something that will take place later within this book. However, at this initial juncture it may be helpful to offer an example watermark that can be clearly heard on the material of the Irish band The Corrs. Robert John "Mutt" Lange's influence is very self-evident on the album *In Blue* (2000), which he not only produced but also wrote/cowrote several songs. In comparison to their earlier album *Forgiven Not Forgotten* produced by David Foster (1995), *In Blue* has a much more commercial pop sound, an attribute arguably created by "Mutt" Lange's watermark.

The producer's watermark is something that will endure with the material. As previously mentioned, it is almost their calling card, but analyzing the watermark more closely shows that this is an intrinsic set of skills, or attitudes to differing situations. Each musician, each song, each arrangement, and each studio will have incredible influence on the outcome. The mastery of the producer in every situation is key and is thus the creative backbone of this book. To begin with we need to get a handle on the types of producer out there.

Production is a very multi-faceted, yet critical, role. The reason why our prospective undergraduate students found it hard to answer the question "what is production" is because it is so hard to define and even harder to write a job description for.

This is our inspiration. Throughout this book we will look toward the process and roles the producer has taken on to get the gig or complete the gig. There is a lot of glamour and intrigue that surrounds the role, attracting many to its doors. However, it is not for the faint-hearted and the responsibility is immense once over the threshold. Once through the door, though, it can be the best job in the world.

Broad types of producer background

The producer's creative input can manifest itself in many forms, from getting down to the nuts and bolts with the musicians in a rehearsal room carving the music into a ready form for tracking, to taking an engineering focused approach, using the technology and the studio (and studio time) to develop the artist's material. Burgess, R.J. (2005) accurately highlights a number of types of producer. Burgess described these types as *songwriter producer, music lover producer* and *engineer producer*. These are fantastic classifications and ones we employ to some degree here. However, for the purposes of this book and our teaching we boil this down yet further to two main simple types: musician producer or engineer producer.

We have taken the liberty of rolling the music lover producer into one of the other camps, dependent on the skills that producer picks up on the job. For example, Burgess makes the case that there are those producers who come to the fray through loving the music and what they'd like to hear. We suggest that many music lover producers will develop either a strong sense of musicality or their engineering skill and knowledge.

Whether their skills develop in either area, it is common for the quality focus of the producer to be honed in either on the music and the musician or more biased to the engineering. An example of the musician producer would be Trevor Horn, while a good example of the engineer producer would be Hugh Padgham, as suggested by Burgess.

Hugh Padgham's career blossomed in the 1980s working with artists such as The Police, Genesis, and their solo counterparts Phil Collins and Sting, among many others. His engineering skills are profound to many and has led to his work being widely analyzed. *Synchronicity* by The Police is one such record. Phil Collins's legendary gated reverb drum sound is also Hugh's achievement, or stumble, or even fault, as many would disingenuously blame these days. How wrong! To Padgham's credit, this was immensely popular, and broadly imitated at the time and this has its rightful place in production history. This trademarking of Phil Collins' drum sound arguably provided Padgham with one of his professional watermarks.

The type of producer you are, if you can indeed categorize, will of course suggest areas of concern for your work. Hugh Padgham during the 1980s and 1990s had the fortune to work with exceptional musicians who had strong plans for their music, support from their labels, and could in turn work with an equally exceptional engineer/producer to gel the team. Padgham slotted into this role thoroughly, producing a raft of top charting albums which to his credit still sound great to many.

It is therefore natural to argue that a musician producer might spend much more time concentrating on the musical development of the artist and subcontract the engineering to a trusted partner they've worked with before. Looking at the long tenure of producer Trevor Horn, he has employed a whole raft of exceptional engineering talent over the years, leaving him to concentrate on the music as a musician producer predominately.

One such intern was Robert Orton. Now an independent mix engineer, Orton spent nearly a decade with Horn as part of his engineering team working on a wide range of chart-topping albums. His work was defined in the engineering and mixing of the productions leaving Horn to concentrate on the musical development in terms of composition, arrangement, instrumentation, and textures he's so renowned for.

Orton is somewhat different insofar as he is an engineer with a very solid musical background. Orton, a pianist in his own right while also being a trained sound engineer and music technologist, bridges the gap between the engineer producer and the musician producer when he takes that mantle. Nevertheless, his role as an engineer while working with Horn required him to focus on being creative with technology whilst leaving Trevor to work his creative magic on the music itself.

Completing your A-team

Whether an engineer producer or a musician producer, the team you surround yourself with will enable you to succeed, or perhaps fail. George Martin had access to world-class songwriters, musicians, and all of Abbey Road's engineering staff (the likes of Norman Smith, Geoff Emerick, Ken Scott, et al.). Would the Beatles have been as successful had Martin decided for whatever reason to have somehow independently recorded the projects himself in a period when independent studios were still relatively rare? Would Hugh Padgham have made a success of so many of those large scale '80s artists had it not been for the production circle his role completed with the likes of Sting, Phil Collins, et al.?

As we will discuss later in Section B, the network of people you can draw upon will become the free-flowing team you'll call upon in production situations. Using your favorite engineer or drummer for a production, whatever the material, is very likely based on experience and perhaps social currency.

Therefore, skills in developing relationships, teams and communities of practice can be invaluable for the success of producers. In this book we'll also look at these aspects and consider how important it is to develop these connections.

Essentially there is no need to state your intent at this stage as to whether your strengths are in musical aspects or engineering ones. Begin to consider the network of people around you who will complete your team, your circle, and help you in the areas where you're less experienced.

Production outlook

The descriptions above discuss how the professional focus of the producer is revealed. However, it is how they apply their art that is of interest. Their outlook on music production is something to consider at each turn and, as with all interactions with other human beings, we're constantly learning. The studio is somewhat of a melting pot and can often bring out new skills and sensitivities in producers. Later in the book we discuss how to integrate with people in the studio and how to get the best out of the production team, the artist, and the musicians around you.

Some producers take a very hands-off approach to their work, leaving their artists to crack on with the music at hand, perhaps believing that the artists may need less intimate direction than other less experienced artists. Other producers may choose to guide and be the fifth member of the band throughout. How the producer integrates with the artist will neatly depend on the type of artists, the level of additional support they may need or the type of music itself. Reading a situation or scenario so as to interpret the level of intrusion or impact required is a significant talent that must be a part of every producer's skill set.

In preparation for this book, we have heard from some producers that have described their involvement in some projects as being very minimal. There are, however, producers who we know have taken quite a leading role in the shaping of the final music, producing its every essence. It all depends on what is required and what the terms of the agreement may be.

Naturally it is very hard to make any one-size-fits-all statements, only to identify the common denominators for many of the producers we've spoken to and we'll leave it for you to develop your toolkit and your outlook. We have taken the decision not to use quotes from all of our interviewees for this book but to analyze common threads, thoughts, and outlooks shared by the vast majority at different stages of their careers and in different genres. It's their outlook that counts, not the anecdotes for us.

Your outlook will be based considerably on the type of producer you are, whether you get on easily with people and can be cool and calm in times of high stress. Developing this outlook will assist in your development as a producer. The ways in which you communicate and describe things will be of importance later on.

Backdrop

Historically, the role of the producer has been that of someone who has managed the project from the finances, studio personnel, musicians, and the product as a whole. Production (for many a traditional producer) was the management of the session and as such typically not always the glamorous role that many associate with production. It could be argued that the process was much more restrictive both creatively and financially. Allied to this, unions such as the Musician's Union in the U.K. then stipulated that all musicians be paid in three-hour blocks. For this reason many sessions would either be three or six hours long. The fable has it that once the three-hour session was over, musicians would stand up and leave mid-song unless another three hours could be paid (how things have changed in rock and roll at least)!

Many musicians, especially some session musicians, who are called in stick to such M.U. rules. Expect this from orchestras and quartets and the like.

In early days of recording many sessions were seemingly limited to this time-frame, based on the belief that nobody really needed more. The aim of the game was to capture a well-rehearsed performance. The production would have taken place beforehand in rehearsal (pre-production as we now know it; see Section C). Added to this, the expectation of highly polished productions using technology was not the norm. Well-recorded, honed, and well-performed, yes, but produced a little less than by today's technologically intensive standards. One could argue that the polished productions expected by today's music industry have shifted the focus from this type of pre-production to production and postproduction, where the knowledge and use of technology now plays a greater part in smoothing off the edges after rather than before the record button has played its part.

Today's producer

Today the word *production* appears to allow different connotations or a wider range of activities than it once did. The diversity of musical genres has now thrown aside the traditional model, allowing for people who, quite literally, physically produce the music to be considered as producers. Using the traditional model, an artist would have been the writer/performer and the producer would be the producer. Nowadays artists can blur those lines, becoming co-producers, and the producer can take some part in the songwriting and performance.

Change the music to recent dance genres, such as trance and house music, and the word producer will inevitably mean the writer and the producer combined. The interesting thing here is that, within the trance and house model, much of the music is likely to have been produced alone with little or no human inter-action in terms of someone to physically *produce*, other than perhaps a solo session vocalist.

Additionally, it is interesting to note that many of the compositions that lie behind dance music genres such as trance require as much effort and skill as traditional songwriting. The difference comes in the fact that a song can be written with an acoustic guitar and be developed to the full complement at a later date. Within trance, for example, this might be somewhat complex and inappropriate to do on an acoustic guitar. As such the composition and pro-duction really are seen hand in hand. In other words, one cannot exist with-out the other; some music cannot necessarily exist without the associated technology. This can add to today's blurring of the term "producer" from its historical benchmark.

Danny Cope, author of *Righting Wrongs in Writing Songs* (2009) suggests "Gone are the days, in certain genres at least, where the creative type needs a producer to make them 'sound good'. That's because of the tools that are so readily acces-sible (Logic, Reason, GarageBand, etc.) which often make it easier to 'sound' good before you have actually composed anything of any real substance. The process has shifted so that instead of writing something and then making it sound good, we have something that sounds good that we then need to create something with. It's like buying the expensive custom-built picture frame first, and then having to paint an exquisite painting to fill it."

Needless to say, the historical production role still remains to some extent. There are many albums created using that traditional production role, but the pres-sures are very different now, having changed subtly bit by bit over the years. For this book we'll continue to focus on the traditional role and from there how it has evolved since. Musicians performing as artists, bands, and ensembles will be with us forever and while they record singles and albums, the objectivity and assistance a music producer provides will always be required.

The lifestyle of the producer these days has changed too from those early days where you once belonged to a record label or recording studio. Nowadays the role is very much based on being freelance and working hard to generate business, thus often with long hours and great personal responsibility. The rose-tinted vision of the producer being in control of all and sundry, firing orders from the plush leather sofa at the back of the studio with a cigar is quite far from the truth. Producers, in our experience, are hard-working, innovative (in the studio *and* in business) affable and entertaining people with plenty to say about their love of music and the way in which it can be developed.

Their working lifestyle is something we'll approach later in this book as we discuss navigating the freelance role and the ups and downs of working in the studio.

The recent climate

The current music industry is a very competitive environment financially, yet there is more music available than ever before. This is mainly due to the widespread availability of music production equipment in the past decade or so. The average home computer is able to develop high, often studio-quality, productions with the correct know-how. That know-how is the all-important aspect and something we encourage people to experience and brush up on. This book describes some of the things it takes to be a producer.

This financial climate is squeezed at the top end and as a result the artist development that was so important to the sound and identity of so many bands has been seemingly reduced in many labels. Artist development whether from the label, artist manager, or elsewhere has been paramount to artist success.

To say that artist development does not carry on inside the labels would be disingenuous, as many artists do of course get the treatment they need. However, the prevalence, we suspect from what we have learned, has been far reduced. The supposed hedonistic excesses of the '70s and '80s within the rock and pop scenes have ceased to exist as we knew them. During this period, very large budgets indeed were poured on productions in the relatively safe knowledge that many artists would recoup the money from sales.

It could be said that the industry was then in a healthy state. Producers were often given complete scope to mold and produce their artist's music as they saw fit with some intrusion from the labels. Some seminal albums discussed within this text were produced in this period.

Among the heavyweights on the rock scene in the '70s and '80s was Queen, a band best-known for their musical prowess, lush vocal arrangements, and unique sound. As producer, Roy Thomas Baker certainly had a daunting task of steering and molding Queen's creativity in the studio. However, time and money was more abundant compared to the climate of today's music industry and therefore spending many days just recording Brian May's multiple guitar parts was often a reality.

This luxury of *time* in the studio environment enabled Roy Thomas Baker and the members of Queen to experiment with different recording techniques and sounds and produce anthemic rock tracks such as "Killer Queen," "Don't Stop Me Now," and the monumental "Bohemian Rhapsody." It is interesting to consider whether such tracks could be produced within today's time and budget-conscious climate, even with the speed that advanced technology affords us.

However, over the years and more recently, the record companies have suffered some reduction in revenue through the widespread loss of sales to digital downloads and digital copying, a topic we investigate a little in Section C. This, in turn, has changed the way in which artists are developed and delivered to the marketplace. It is not uncommon for a good quality demo to be placed on the shelves with some minor changes. The aforementioned album from David Gray, *White Ladder*, is a fine example of this. The demos were rebalanced and all the vocals rerecorded, but the organic package was too precious to alter. This strategy proved to be correct, making the single "Babylon" a Top 5 hit in the UK.

A slightly different but interesting situation occurred in the U.S. in the mid-'90s with the band Jars of Clay, who while playing together at college submitted a demo recording to a talent competition. Upon winning they were given the opportunity to play in Nashville. The buzz from the performance coupled with the popularity of their original demo saw record deals being offered from labels. Further down the line the demos were rerecorded in the studio for their first album (self-titled *Jars of Clay*) and the band have since gone on to be multimillion selling artists with two Grammys to their name.

These changes in the current music industry climate have led to many other avenues or types of signing. Many genres have sprung up and remain quite small in terms of sales and clientele, yet have sustained a solid following for many years. Such diversity has only been prevalent for a number of years now and offers musicians and aspiring producers the opportunity to get their music heard.

In conjunction with new websites offering distribution, it has never been easier to have a professional presence without a recording contract *per se* and reach the public. Other methods have allowed for many bands to continue with their careers, without the machinery of big labels. One such example is the U.K. band Marillion. They have used many different methods in recent years to enable them to write and produce their next album. *Anoraknophobia* in 2000–2001 was one example. In 2000, without a contract or the money to release a physical album themselves, they decided to use other methods. The one big leap for them was to email their fan base via the website to ask them if they would be prepared to pay for an album up-front. This was very successful and enabled them to make a record for their fans, paid for by their fans.

Although not a new method by far, a recent example of DIY can be seen in the career of folk musician and songwriter Seth Lakeman. Lakeman's self-written and produced solo album *Kitty Jay* was allegedly recorded at home for just £300. Lakeman is now signed to the record label Relentless and has received much critical acclaim and success.

Another facet of the technological explosion is that producers of a different variety have emerged. For example, despite rap being entrenched in popular music culture for quite a while since the 1980s, its now enormous mainstream emergence in the U.K. was given a boost by Dr Dre and U.S. artists such as Eminem. Dr Dre is both an excellent producer and businessman, having brought many major selling acts to the fore. Within this genre, it is important to demonstrate the right sound, impact and base upon which the rap resides. Dre mastered this to a fine art.

To compare radically differing people such as George Martin to Dr Dre would reveal some interesting disparities. Not simply their upbringing, education and musical tastes would offer differences, but the management of their artists would be completely different, as would their business dealings.

It would therefore be fair, and obvious, to suggest that differing musical genres require different treatment, and in most cases a different producer and production team. There have been few producers that have spanned a wide variety of genres. One example is Robert John "Mutt" Lange, who has worked with pop act The Boomtown Rats, to soul/pop Billy Ocean all the way through to the heavier rock of AC/DC and Def Leppard and more lately based within country/pop, producing his former wife Shania Twain.

Throughout this book we will try to give examples of the industry as it is now and how this affects the role of the producer. However, we are mindful that the industry moves so fast and that some examples will be less relevant than others. Therefore, we have selected those examples that we believe demonstrate the discussed matter well and thus the information can be transferred to a wide range of situations. We have placed sources of interesting comment and further reading in Appendix F-1 –the tape store.

UNDERSTANDING THE PROCESS
Understanding what is required

It is always interesting to read interviews with the major producers in the trade magazines. They are portrayed often as having planned out the work they were producing. However, this is unlikely to be the case every single time. Only so much preparation can be completed prior to recording. So many artists now write in the smaller studio and as such the production happens there and then as an iterative and integral process. This is of course so much easier by today's standards now that we have nonlinear editing and all the gadgetry.

However, planning is something everyone can improve upon, and in many a music production it can be vital. We spend some time in Sections B and C looking at the behind-the-scenes work of a music production and the preparation you may need to consider.

In earlier times, producers had only preparation to guide them to a successful conclusion. Often limited by the three-hour block of time, the producer would use an arsenal of skills to prepare the music for the session. This understanding gives way to arrangement organization. Phil Spector is regarded as one of the masters of this with his work in the '60s. He would place musicians around a large live room to create a large, almost orchestral, "wall" of sound.

A well-known example of this painstaking arrangement methodology is "River Deep, Mountain High" by Ike and Tina Turner (written by Jeff Barry and Ellie Greenwich). The large ambient wall of sound that can be heard in this track and many others not only became Spector's production watermark but also a trademark of the era. Incidentally it is documented that Spector worked Tina Turner hard during the recording sessions in order to achieve the desired energetic and powerful vocal delivery. This attention to detail was based around what we refer to as the producer's CAP (Capture, Arrangement, Performance), which is covered in Chapter C-4.

The brief

In so many assignments, a brief of sorts may be given or directed by the label's Artist and Repertoire (A&R) department (see later in this section for more information). Somewhat inaccurate folklore would have it that the A&R department will throw in a brief something along the lines of "This band needs to sound a little like Oasis, with lots of Orchestral Manoeuvres in the Dark, with a splash of Norah Jones thrown in during the choruses." Perhaps this is a bit too flimsy and diverse an example on our part (and with no offense to the artists who are incidentally held in good regard by the authors). We agree that it is an extreme example of the directions you may be pulled in as a producer. Realistically, of course, you will be able to discuss a better and more suitable alternative route.

Your incentive may be that "you are only as good as your last record." This is all you have to go on. The manipulations from either the A&R or the band may conflict with your ultimate plan for the band or artist in question; you will try to reconcile this with your ultimate plan and direction for the band, while making great music, which is, after all, what you're in the business to do.

This conflict may happen internally for the most part and thus cause a lot of deliberation. However, a conclusion is required, and therefore it is often necessary to be rather visionary at this stage. It is important to bear in mind that not all productions require over-production. Sometimes, as with a mastering engineer, it is often best to live by the adage that "less is more" or better still "know when to do nothing at all."

Many producers, we often think meddle with the music presented to them, not because it needs it but because they feel that they need to be seen to have produced it. In their eyes, this means altering what is there. One producer we spoke to was amazed when he was asked to work on a band's next album because of what he didn't tell the band to do!

In this instance the band recognized that the production sage might opt to alter many minor elements and that was critical, meaning the producer maintained the rights to the production credit. This less-is-more approach is critical to success.

There are so many artists out there who simply do not require the high production so often applied to them. They can carry their own. Even Peter Gabriel in 2010 called in Bob Ezrin (producer for many artists including Pink Floyd) to assist him in the production of *Scratch My Back*—not as the producer, but as almost a production consultant. In this instance Ezrin's brief will have been to support and assist the development of the album.

Rupert Hine, while speaking with us, discussed how he has started working in a way that could be described as a production consultant. He now works with some artists remotely. The artists will send Hine MP3s of the musical development and he can then reflect and discuss with the artist, steering the musical outcome.

Artist & Repertoire?

A&R have received a bad reputation unfortunately, not least because the tradition of A&R is said to be lost. These departments were once responsible for bringing the artists together with their songwriters and producing a format and a buzz. For example, Frank Sinatra did not write his most notable material. It would be partly the duty of the label to source material and songwriters for artists and match them together accordingly.

So what changed? Well, as artists began to write more of their own material (the Beatles being an excellent example), the A&R person developed to survive. In doing so, they continued to act as the talent scout but also, as Neil Peart in his 2004 book *Traveling Music* puts it cynically, A&R may meddle with the music based on the label head's guidance. "[It's] the kind of meddling of the 'business' in the 'music' that always bugs me."

But let's spare a thought for our colleagues in A&R. Their careers are insecure with high turnovers of staff. Again the adage "you're only as good as your last record" (or in this instance, "signing") is applicable here. Consistency is lacking and as a result reputation and the artist development has been under some jeopardy.

It is widely believed that A&R meddle in mixes and final output based on a whim, or a musically inept perception of market influence at the time, which may in turn jeopardize the music they are attempting to promote. Therefore the role of the A&R department is often understated and undervalued within the marketplace by both employers and the music fraternity alike. A great shame, perhaps? We need to look far more closely at the issues surrounding the record labels and the current investment levels in new acts before we should judge too rashly.

The term *production*

"Production" can be used to describe so many different things these days. In this book we discuss production in two ways:

First is what we would naturally refer to as *music production*, which is an expression of the creative and artistic development of music both in and out of the studio. While this book does not intend to address the whole gamut or any specific creative and artistic elements of music production, it does deal with the back-office work and the day-to-day life that the producer undertakes.

Second is the *production process* itself. By production, in this latter instance, we refer to the process of producing an end product: the making of the CD; the making of the media that sits on a server for download through an online store; the physical medium perhaps not yet decided upon which will convey our music to the masses in the future, if a physical medium is even needed at that point.

This latter reference to production is the process we outline within this chapter. Each stage of the process has its own innate history and has developed into a well-oiled machine that responds with innovation to change and style throughout time.

The production process

Anything published, whether a book, a magazine, a movie on a DVD, a website, or an audio CD, must follow some kind of *production process*. In this instance, we do not mean music production *per se*, but some kind of production process, as defined above (Blu-Ray? SACD?, etc.).

This production process has, for whatever artistic output, been carved out from initial trial and error, and through painstaking refinement, to create and define the present systems and procedure we rely upon today. Any artistic process can follow a generic model, similar to the one shown here. In this example, we've identified seven rough stages of the production process.

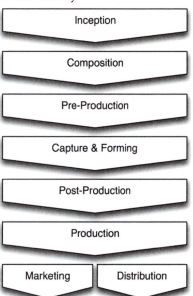

Any creative, yet commercial, process resulting in a final product can follow a model similar to this. It is worth noting that this is by no means exhaustive and variations can be the norm.

To produce a creative work, an idea must flourish, or the desire to capture an experience. So often, many of these ideas fall by the wayside, but from time to time, they will make it to the drawing board. The composition of the music is mainly outside the scope of this book as we make the presumption that, as a producer, you'll be working with an artist. However, it must be pointed out that in this day and age of 360° deals, producers are having to protect themselves and their income streams. As a result, the clever and insightful are beginning to engage much more in the writing process. We discuss income streams and 360° deals later in the book.

The next stage in the process is pre-production. In this book, we dwell for some time on this topic, looking at its importance and some approaches that can be employed to improve productivity in the remainder of the process. Preparation for any creative project can seem to detract from the art form, but in the case of any production, there is a business side that needs to be considered and supported.

It is imperative that planning is given equal value, or equal consideration, when contemplating a project. Is the project viable? How should the project begin? Who should be involved? These are all valid questions that require valid answers before commencement (or sometimes during): Answers that will inform the rest of the process and the ultimate success, both financially and artistically.

As with any art, it needs to be captured so that it can be portrayed. Paintings need a canvas upon which they can be structured and later viewed. Music requires some form of canvas too. This can of course be manuscript using traditional notation, or a sound wave captured in either digital or analog form. The captured work needs to be structured or formed. In music production and recording terms, forming means mixing so the elements can be balanced accordingly. For the painter, this would mean less red, more light, and so on.

Postproduction as a stage in the process is the first stop on the mass-produced train track. For a painting to be mass-produced, or copied, it must be encapsulated first. Once copied it can then be prepared for mass-production. Any edits in light, color, or shade can be applied to ensure a maintained quality across the various mediums (postcards, posters, prints, etc.). It is the same for music. Music needs to be prepared for the medium upon which it will be presented, or mass-produced. Additionally the quality can be improved, or balanced, at this stage.

Next in the process is the production itself—the way in which the mass-produced material is collated and reproduced. Replication, distribution, and marketing are things that are of least interest to the artist compared to the inception, composition, and capture of the art. However, it is imperative for the mass-produced reproduction to retain as much quality as possible. Equal interest should be paid to this part of the process (as it is in this book).

Traditional roles in the studio

The roles of the production process have remained broadly the same since the inception of the recording studio. Naturally much has changed over the years due to the necessity of some maturing in the process, but also because of the advancement in technology and the ability to bring so many roles into one should that be necessary. Some artists and producers still have a strong preference to have a larger team than simply one engineer/producer.

Historically, sessions were attended by a small group of specific roles with large responsibilities in each department. The sessions would be managed by the producer, and in attendance would be an engineer, an assistant engineer and/or a tape operator. It would even be possible on some larger sessions for more than one engineer or assistant engineers to be present. The equipment might, of course, require this, as the tape machine should be attended to regularly and managed. One slip-up here could cause loss of that perfect take, so to manage this at the same time as a mixing console would (in a time before auto-location and memories, etc.) be perhaps too much. As such the division of roles was very structured and simple to comprehend.

These structured roles have been used in our teachings for many years, as they allow not only a sensible division of labor within the control room, but also engender sensible educational group work—something that is so crucial in this industry. The advent of the computer, with its own onboard mixer, has naturally blurred these strict lines. No longer is there one mixer in some studios, but two. For example, Robert Orton chooses to mix completely "in the box" within Pro Tools, simply resigning the SSL on most occasions to no more than a stereo volume control via two faders (a 2-track return essentially!).

Recording sessions these days can be managed by one person if desired. As previously mentioned, modern studio equipment allows the engineer alone to manage and develop the session if required. Therefore three or four engineers are no longer required in each session, but just the one if necessary. Additionally, this person could be the engineer and producer rolled into one.

Therefore, the roles in the studio will alter dependent on the session ahead. A simple acoustic recording or rock band perhaps can be managed by one person alone, while an orchestral session will require you to engage a number of assistants to simply move between the control room and the live room to alter microphones.

Historically the producer would, as we will explore in this book, be in charge of the production process, timings, and artistic direction. The role encompasses the financial side of the production in addition to all the nitty gritty project management we'll explore later.

The engineer, meanwhile, would preside over the management of the audio signal through the console through to the multitrack machine, ensuring it is tracked appropriately. In addition, the engineer would ensure that the session runs smoothly from the perspective of the engineering team. Alongside the engineer would be the assistant engineer, ensuring that the microphones are placed accordingly and that the musicians were helped in achieving the sound required. Naturally the role also would be to help track signals if required, amongst a whole plethora of other activities.

It would not be uncommon on larger sessions for a tape operator (tape op) to be present. The role of the tape op would be to manage the multitrack, ensuring that the tape is aligned to the heads, the tape machine is cleaned, and may ensure also that track sheets, take sheets, and other information about the session are stored appropriately.

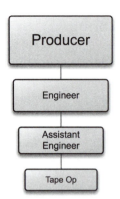

The traditional hierarchy in the production team within the studio. Additional roles can be appended as required, or indeed removed.

Tape ops have often had some of the 'lesser' jobs in the recording studio, hence nicknames such as "tea op" for their ability to make the all-important 'cup of tea' (possibly coffee in the US!) during sessions. Their job descriptions have often left a lot to be desired, from having to work on reception overnight, through to providing errands for the band and engineering team, to gathering various uncommon supplies, that some consider essential to the smooth running of the session. (This has perhaps more recently been replaced by the studio 'runner'.)

Climbing the ladder

Getting on the traditional studio structure was an apprenticeship style route where a budding engineer might begin as a tape operator if lucky, or just as a member of administration staff such as receptionist. When an opening to operate the tape machine came, the young engineer would step up to the role. Many stories emanate from the studio of assistant engineers or engineers either not showing up to work or being ill, leading to a hierarchy shift where the assistant engineer would step up to the engineer mantle, while the tape operator would step up to assistant.

This folklore, while very true, is less the case these days given the aforementioned changes in technology and the personnel structures that surround it, but also due to the reduction of studio space these days. Assistant engineers are still very much part and parcel of the larger studio spaces.

The times are changing and records are made in new interesting ways these days. However, in the major recording studios in the U.K., this is still a typical career structure for some. Those who embark on this route into the career are inducted and are trained thoroughly in their activities and are able to network throughout the industry from the very first spell on reception.

Others perhaps come into the career at differing points of the hierarchy. Robert Orton came to prominence via a different route. He started working at London's Phoenix Studios, then in the former CTS studio building in

Wembley, as someone who would be around the studio to help and assist on the larger sessions. He made sure he learned as much as he could about Pro Tools, looking over the operator's shoulder and picking up all the tricks he could!

Regrettably as Phoenix was given its orders to move out to make way for the new Wembley Stadium development, Orton looked for new work and landed on his feet, becoming part of producer Trevor Horn's team at Sarm Studios. Early days would see him managing Horn's Pro Tools rigs, backing up files, and managing the systems. This led very quickly into Orton running some sessions on the computer and having a go at a mix which Horn liked. In partnership sometimes with other engineers at Sarm, he shared mixing duties on albums for artists such as Seal, Pet Shop Boys, Robbie Williams, Captain, Tatu, and Enrique, among many others. Since leaving Sarm he has paired up with producer RedOne and has mixed for many current artists such as their Grammy award-winning work with Lady GaGa.

Orton has a unique experience where, still relatively young, he has achieved his perfect role as a freelance mix engineer. Orton is lucky, as many still work in assisting roles in the main studios for many years before gaining the opportunity to move up the ladder.

Some of course transcend the ladder and enter the role of the producer at very different places. Some become producers from the mainly musical route, such as Brian Eno, Trevor Horn, Rupert Hine, and Bernard Butler, to name just a few. However, we must point out their personal fascination with what technology could offer them!

Their route is often a less secure transition somehow. There have been many successful artists who have tried to move over to the role of producer and have not quite managed to do so in the mainstream. Other people, through self-producing, are then given the opportunity to produce artists based on their own personal success and this has become a more common route as time goes on.

However, the true development for the future will be the producer's ability not only to manage the production process and the music, but also their ability to find, develop, and write with the artist in a longer term arrangement. Many producers are understanding that this arrangement is a more certain way of receiving an income from the work placed in, as both advances from labels and record sales are actually meaning a reduction in real terms in income.

This arrangement will allow, more than ever perhaps, the opportunity for budding producers and engineers to make a go of it themselves in the absence of a supportive and powerful culture of record company support and development. The do-it-yourself (DIY) opportunities now afforded to producers through the Internet are rife and will perhaps become the industry of the future. We discuss this later in Section C.

Day to day

For former Abbey Road Studios engineer Haydn Bendall, music production is a fabulous job in which there's something new every day. "It's not as though I'm pulling myself into a job I do not want to do for eight hours a day".

For many professionals, work is simply work. It's something that happens at work and does not come home with them or to the pub or bar, for that matter. However, music is constantly all around us and there are times when one's sound engineer or music producer ear kicks in involuntarily. The ear unwittingly digests or analyzes what is being heard, which is not something you always want when you'd rather simply appreciate, admire, and envelop yourself in some favorite music.

Life as a music professional can be different and, to friends and family, a little odd sometimes, especially when you take exception to something you hear. Here is a warning that can be disseminated to music production students: you're always listening!

The role of the producer is sometimes framed as being the cigar-smoking wise sage sitting on the sofa at the back of the control room. However, this is a misleading image. Producers are very often grafters, or at least their teams graft with them and on their behalf, and are a truly dedicated breed committed to good, innovative music. How each producer operates will differ between individuals, but the dedication to great-sounding music is key.

Suffice it to say, the role is not all a bed of roses. There will be times when despair and frustration are normal. Difficult artists can become hard to manage in the studio and the sensitive flux that holds the productive environment together can be shattered in one less considered comment.

Producers never cease to amaze us in that they learn to transcend many of these issues. Some producers we have interviewed for this book are some of the most humble people we know, yet are also at the same time quite opinionated about musical direction. Their easy manner and honest dedication to the music gains them appreciation from the artists they work with and can often ensure the session continues smoothly.

This delicate balance between the flux of the session is something the producer becomes a master of on a daily basis. Ensuring this remains a productive and creative balancing act will result in a free-flowing recording session. We'll cover how to develop these ideas and skills a little later in the book.

The day-to-day role of the producer is in no way prescribed and can be very varied from a meeting with a label, to meeting a new artist at a gig, or sitting with a colleague in a studio editing some vocal takes. Each day can be different and rewarding. However, at the same time editing dull and boring vocal takes can

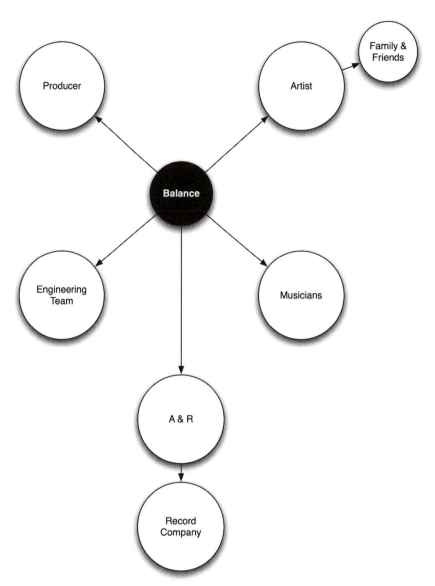

The balance of opinion, or what we call flux, is key to success and expands upon the triangle of influence—something the producer becomes a master at managing on a daily basis.

become a thankless task too. As with any role in life, there are the good bits in addition to the bad bits. The producer has some really exciting parts, and a lot of more mundane parts that are hidden from the public eye. It is these we wish to unearth in the remainder of this book.

"We're not that important in the big scheme of things" says Haydn Bendall of the music producer's role.

Analyzing It

FORMING AN OPINION

As with many art forms, critical analysis is a way of life. To form a view of another artist's work is not only necessary for personal professional development (learning, listening, developing a personal 'watermark'), but also for producing future artists and positioning them in their particular place in the market.

To be appreciative yet critical of colleagues' work is a necessary evil of becoming a practitioner. When we say "critical," we do not mean a type of horrid negative critical disapproval, but an academic and professional critical approach. This could be conceived as a method by which you benchmark the music or audio in front of you, or in which you can break down the construction for future use in your own work. Some critical analysis will be not simply of the music and its construction, but also social and cultural aspects too: how the music fits in the genre, how the music represents the times, and so on.

It can be difficult to form a proper opinion of a recorded piece of music. You'll note the addition of the word "proper" in the sentence before. This is where the difficulty lies. It is very easy, in fact natural, to either be attracted to something or be slightly repelled by something, whether that is a fellow human being, a piece of art, or a music production, for many unknown reasons.

These opinions are partly what form us as individuals and allow us to have engrossing discussions on our chosen subject matters. However, what is a proper opinion? Why do we feel attracted to one person, or another; one piece of music or another? Some things just press the right buttons, irrationally or rationally, consciously or subconsciously.

It is important that emotions should not dominate us, but guide us. Objectivity is a powerful and desirable attribute for people of all positions and fields. Understanding the point of view of others and their artistic philosophies is a necessary, some might say crucial, skill.

Balancing a subjective opinion with an objective one can be extremely helpful and something we all must learn to manage. For example, many producers will speak of creating a *vibe* in the music, or extracting this aspect of a recording to

What is Music Production. DOI: 10.1016/B978-0-240-81126-0.00002-0

enhance its immediate attraction. This *feeling* might be based on the subjective and personal emotional response of the producer. It is natural for musicians to go with this feeling, run away with it if you like, and allow this to guide the session. However, it is the *objective* view that might prevail to place a reality check on proceedings, placing the material back into current trends and the marketplace.

In some ways, the subjective and objective views can cross in that you may objectively allow yourself the opportunity to subjectively be moved by the music (as though a member of the buying public) by means of a test. Does this dance track make me dance? Does this metal track make me want to rock my head? Does this chill-out track make me chill?

These are all fairly subjective and sometimes irrational responses to music, but these things move us. They make us tap our feet, turn up the stereo, drive a little too fast, nod our head, and so on. Music simply moves us. These are the sentiments and the triggers, often unexplained, which make us become musicians, producers, or simply music lovers.

This fascination raises all kinds of questions which we'll not attempt to address here now, only that to be able to switch this on and off can assist you as a producer. An *emotional listening switch* if you like.

Being able to control these emotions for your advantage as a producer may become something that is of great value as a skill or commodity. To be able to listen as though you were a member of the buying public (music-lover), while having knowledge that what you hear in that state can feed directly back to you as a producer, in some ways provides instant market research.

Music Lover

Music Professional

**Emotional
Listening Switch**

Wouldn't it be useful if we could tap into some instant objectivity, or instant market research, by employing a personal emotional listening switch? No longer would we listen as a default industry professional, but perhaps as a music-loving member of the buying public.

Many might say that forming an opinion based on what you like, or what moves you, is very subjective and could be misguided. It could be argued that successful business can take place only with a cool, calm, shrewd look at the marketplace and with good management. This appears unemotional, as in many businesses, art is not at the core of day-to-day trading.

Music is rather strange in this regard, certainly the world of music production. By its very name, it is making music a product, therefore a saleable item. It automatically combines art form with marketplace: something produced by love, later sold for commercial benefit.

The picture painted above almost suggests that art sold for financial gain goes against the grain. However, this marriage of art for commercial success has become the trademark for many producers.

Stock, Aitken, and Waterman spent successful years seemingly producing hit after hit for any new artist who came along. It almost did not matter whether the artist could sing, but only whether they could sell records appropriately. That misconception is what many cynically believed. However, this era launched many long-lasting and highly successful careers, most notably Kylie Minogue, already a successful actress in the Australian soap *Neighbours*. There is no doubt from the interviews we've held, that she can sing!

Other producers have been highly successful in marrying the art form of music production to commercial success with assembly-line precision. Trevor Horn could be listed here for succeeding with a wide range of artists over a lengthy career. More recently Mark Ronson has developed a producer's watermark which has been successfully applied to many artists' work ranging from Amy Winehouse to Duran Duran. Many producers have been successful in this regard and across many genres, Robert John "Mutt" Lange being another.

By no means does this apply universally to all their work. There will be the albums you've never heard of or had the pleasure of hearing. Haydn Bendall, engineer and producer who has worked for many leading lights such as Paul McCartney, Kate Bush, George Martin and countless others underlines the often forgotten (in the industry) absolute importance of the artist. He quotes for example, the fact that on one record the producer can do no wrong and produce an amazing album for one artist. With the next artist, using the same mics, studio, desk, and so on the story is completely different. Bendall makes the point that the artist still can and must have a huge sway on the success of a project obviously from the songwriting perspective as well as the performance. The producer can only do so much. The producer's arsenal of abilities will be more popular in some years than others and because of this one will go in and out of fashion. Keeping abreast of current practice is an ongoing pursuit not restricted to simply the music industry.

One skill a producer has, or should have, is the ability to be decisive and form solid, articulated, opinions—some might even say opinionated! Analyzing this rather negative word with all the connotations we expect from everyday life suggests that we should attempt not to be too opinionated and not to impose our thoughts and wishes on others. This obviously is transferable to the world of the music producer, but some etiquette remains. Haydn Bendall asserts "...we're an opinionated bunch. Well, we're paid to be so. To make decisive decisions is our business. If we've got it wrong we'd be the first to admit something is not working, but we like to try it out first."

Therefore, the producer's opinion is of increased value to the production of music. Some artists will require that firm artistic hand to interpret their work. Peter Gabriel writes in the sleeve notes for his 2010 album *Scratch My Back* that he set out some criteria for his arranger to work with. He indicates that to offer an artist a blank sheet can sometimes be the worst kind of freedom. Placing some restrictions can be the way in which we develop to overcome the rules. Gabriel is right and we speak later about serendipity and that restrictions can mean outcomes!

Forming an opinion is an important attribute. How else will you know what to repair, modify, scrap and so on? How we manage our conviction is another matter. We believe it is important, as you will tell from many sections in this book, for a producer to be an exceptional listener while also sympathetic and intuitive to clients' needs.

Form an opinion. It is absolutely necessary, but should be drawn from both subjective and objective frameworks dependent on the task at hand. But be open, objective, and able to admit you're wrong all at the same time. Managing this is something that takes time to master and understand. The following sections relate the ways in which we can listen to music in so many different ways and on so many different levels to give us the edge and an open, objective, and modest viewpoint.

DETACHMENT

So how should you be guided through an opinion? By your gut feeling or by some rational objective view? Well, this might be up to you. There is a detached way of listening that is good to try to develop. When listening to music, it can be difficult to detach yourself from the subjective, enjoyable aspect of the song you're listening to (the head nodding, the foot tapping). After all, this is the reason we're all in this, because we love music. So what are we talking about? It can be difficult to form a professional detached opinion of a piece of music you grew up with, or feel is somewhat second nature to you as a member of the music buying public.

Your reason for enjoying this piece is because it was something special to you at the time, perhaps even employing a hairbrush as a microphone! You were perhaps not a music professional then. In any case, the music would be something you appreciate for its art, its style, and simply to enjoy. As the music professional within you grows, you can begin to develop the ability to listen outside of that enjoyment bubble that the track generates and analyze the piece accordingly for all its merits and flaws.

We see this as a kind of switch that can be turned on or off as necessary. However, many music professionals report that they have had periods of finding this switch hard. One engineer we spoke to said that he could not enjoy music for a number of years as he would always be in analytical mode constantly and unable to let go.

Obtaining this switching ability is extremely helpful, as it assists the music professional to engage in the two types of listening, from the enjoyment aspect and from the detailed aspect, providing it can be controlled.

It is not simply about detachment. There will be some times when you may have to work on material that you do not enjoy, that is perhaps out of your comfort zone. This is where a detached, objective ability to listen and react will be of great benefit.

Within this chapter we investigate some ways of listening and how to begin to delve into the world of analyzing already produced music.

UNDERSTANDING LISTENING

Let's state the obvious here for just a second: listening skills are an integral part of the music producer's, indeed the music professional's, portfolio. However, it is, as always, not what you've got necessarily, but how you use it.

If we were to canvass all the producers and engineers in the world, we are sure many of them would not profess to have perfect hearing by any stretch of the imagination (listening for long periods at high SPLs can take its toll!). Neither would they all profess to have perfect pitch, because they don't.

However, they would all claim to be able to listen. Listen intently. Listen accurately. They might also claim to be able to hear how they wish things to sound before they've heard them. Therefore it is what we refer to as *ways of listening* that interest us later in this chapter. Some are perhaps born with the skill to be able to hear and listen in a detailed way. However, many of us have to train ourselves in the ways of listening.

There are many websites and products offering to teach you methods of how to listen in different and new ways similar in nature to ear training software, such as Auralia, which focuses on identifying pitch and chords within music. However, we've found one of the best ways of learning to listen is to actually *listen*. If you're a new producer or engineer just starting out, simply buy a trusted pair of common monitors (in our day this would have been a set of trusty Yamaha NS-10M Studios and to many a pro they still are). Listen to as much music (or everything) through them, set up in as decent an environment as possible: minimal reflective surfaces with the monitors forming an equilateral triangle between themselves and the listener.

The premise behind this is that your ears will develop a transparency, or an affinity, or relationship, with those speakers in a good nearfield environment, the very same environment that you could replicate in any studio you might wish to work in using your own speakers.

The benefit of this is that you will always have a trusted set of monitors with you wherever you go, allowing you to produce results quickly and efficiently (or indeed you can research the monitors used in the studio you most use). You can do this because you've listened to all your reference material on the same speakers (and hence focus on every single element) plus you also know the voicing of the monitors (how certain instruments stand out; for example the NS-10Ms would elevate vocals out of the mix a little).

This can put you *in the zone*, making a transparent connection between your ears and the music (in other words, less concentration is being spent on real-time compensation for monitors you do not know). This in turn should allow you to make more confident decisions without needing to listen as often to any reference material you may have, thus maintaining focus on the music at hand and saving valuable professional time. In these days of high quality active monitors, amplifier selection is less important, but should you be employing a passive set of monitors, then amplifier selection can also be critically important.

Getting your monitoring environment right is also key, and regrettably outside of the scope of this book. Nevertheless, it cannot be stressed enough how important it is to ensure that your listening equipment and environment is as solid as possible. Good monitoring in a good room should provide an excellent listening environment. Remember, assessing music and its components in anything less might skew your thoughts and actions later in your own work.

You may wish to emulate the conditions you find in the studio by completely listening in a nearfield environment. If you have nearfield monitors place them at the same height and orientation you'd have them on a meter bridge, thus experiencing things as they'll be when working hard on the session. Some engineers have spent their careers carrying around all sorts of trusted loudspeakers for their projects. Famously, Bob Clearmountain has for many years employed a set of Apple computer speakers among various sets of mixing monitors with award-winning success!

Now here comes the caveat. This might be a good way to train, especially if you are interested in the engineering side of the business. Many producers will use other systems and environments as their check. Many use their car stereo systems, as these can be one of the most common listening environments despite their poor reproduction characteristics. Lets face it, many producers spend a lot of time in the car between sessions, meetings, and gigs. Really, it is a case of whatever works for you. If there is a motto in this area, it would be to be confident in what you hear.

THE REFERENCE MATERIAL

So what do you do if you're working in a studio with different sounding monitors and have not got your own to hand? Well, you employ your reference material.

It can often be hugely impractical to move your monitoring rig around from studio to studio with you for every project you work on and therefore you need a reference point which can help you be educated to the tone and color of the monitors and acoustic environment in which you're working.

It is popular to carry CDs (or non data-compressed .aiff or .wav files) of a selection of well-known (to you) tracks across a number of genres. These tracks might have aspects that simply work. One track might have an inspirational acoustic drum sound which can often be a point of guidance while tracking or mixing. Another track might demonstrate full dynamic range and also have a fantastic blend of instrumentation; this material might simply be an excellent mix. This excellent mix can then become a blueprint of the relationship to the differing frequencies you may expect in the mix and may choose to emulate. Of course there are many fantastic benchmark albums you could employ for this. Some might choose seminal albums such as Pink Floyd's *Dark Side of the Moon*, Michael Jackson's *Thriller*, or Def Leppard's *Hysteria*.

One thing this reference material should permit you is the ability to discern the characteristics of the new monitor and its environment. On listening to your well-known reference material you might note that the treble is quite exaggerated while the mids are subdued, leaving vocals in some ways less present in the mix. With this assessment on board, you can compensate for your lack of empirical knowledge of the monitors in question.

One way to have this to hand, should you be engineering, is to have the CD (or other player) connected to a two-track input on the mixer so you can easily switch between your mix and your respected reference material. Naturally we ought to bear in mind that the reference material will be mastered, but you should be able to work around this, bearing in mind that your dynamics should be allowed to breathe perhaps a little more than the mastered version is. The frequencies and so on can remain a consistent and reliable guide to your work in the sessions.

These simple, yet time-tested, tools can improve your productivity in the studio as your sonic decisions may count on it, should you choose to work this closely with the engineering side of things.

Many people still confess to mixing on headphones. There are many arguments that suggest this is not such a good idea, but as the buying public are spending more and more time on their earbuds perhaps this is something professionals should consider as routine when checking mixes.

How the record-buying public listen

Music professionals, it is argued, should strive to be confident in what they hear and be able to make appropriate assessments accordingly. We're sure that many professionals would prefer to be using their treasured monitors and listening environments to make such assessments. We must, however, all recognize that the buying public listen to music on a wide variety of equipment, formats and modes.

As such it is surely the goal of any production team to ensure that the music being produced should be intended for the equipment on which it will be played. Well, yes to one degree or another. Dance music should, it therefore follows, be intended for the club PA, while the singer/songwriter's delicate album might be best placed for the earbud headphone.

Cross checks are also required. It is important for both sets of material to translate to as many different listening environments as possible, including the car, and to sit together if segued together on radio side-by-side.

A large debate has begun and continues to rage within professional circles about the way the public listens to music at the current time. There is no dispute that the generic MP3 player or phone has begun to dominate the listening market. It comes with some fairly inexpensive earbud headphones with reasonable reproduction. Some professionals are thus concerned that the listening public may never hear the music in its intended sonic form over loudspeakers and in a high enough resolution.

Meanwhile, the second most popular environment must be the car, which can be rather unpredictable. However, many engineers and producers still ensure that their material works in the car by taking the material out there during mixing, and even mastering, sessions.

Therefore, some might recommend that professionals begin to benchmark their material using the same systems. Using an iPod for monitoring might seem wrong, but might be good to see if the mixes are translating as one might expect. The NS-10s of the future?

DIGITAL AUDIO, IPODS, AND SO ON

The arguments rage on. The music-buying public vote in force that downloads, and the data-compression currently used, are the way they choose to consume music.

Until this point the music lover has had to import her CD into a library and the file format would most likely be MP3 or similar to save space on storage capacity on both the host computer, but also her iPod.

It is rare that the CD would ever be listened to live again and perhaps most likely will have been simply stored. As a result, many people are finding downloads (legal or otherwise) convenient as it can be fast, saves a trip to the shops, and the quality (MP3/M4A) has become acceptable.

File formats generate a great deal of discussion insomuch as they do not deliver the full bandwidth of the music. Some rudimentary calculations would tell you that a basic MP3 stores around a fifth of the data that a CD would, depending on the bit rate. This is achieved by some quite clever data compression. In brief, many data reduction systems work on a range of principles and result in either *lossless* or *lossy* methods (see the Glossary for more information).

One interesting principle is based on the presence of frequencies that we do not concentrate on in the presence of adjacent loud ones. In other words, the frequency spectrum is chopped up into lots of bands. If one of those bands is loud at any one point, it is likely that the two adjacent bands on either side are less important. Using some fairly sophisticated algorithms, the encoder will manage these and perhaps choose to omit them for a number of samples, thus saving the amount of information recorded. Naturally this principle of auditory masking is actually more complicated when employed, and readers should be directed to *The MPEG Handbook* by John Watkinson (see Appendix F-1) for more information.

An argument still exists that CD is not a high enough quality. Whether we like it or not, professionals spent many decades prior to MP3 striving to improve the sound quality and to some the CD allowed a solid delivery medium to the consumer of high-quality reproduction. The CD for the first time omitted the flaws associated with poor styli and turntables where pops and scratches had an analogous effect on the sonic quality. Many simply feel that the resolution of both the sample frequency and bit depth are not enough.

The iPod and its culture have changed this perception. In order to squeeze material onto limited storage capacity or to be able to transmit it across the Internet, the MP3 format was adopted and lately the better sounding MP4/AAC codec.

There are many within the industry who currently hope that the day will come when it will be possible to download full bandwidth material at perhaps higher bit depths and sample frequencies than CD's 44.1/16 (44.1 kHz sample rate and 16-bit word length; see Glossary for more details).

In fact, the case is often argued that a new 192 kHz sample rate and 24-bit depth could be set as the standard for all multitrack audio, thus eliminating any need to look at other rates. This naturally would support high quality audio at every stage even if it did end up at 44.1/16 on CD.

What has transpired is that most engineers will record definitely to 24 bit as standard with either 44.1 kHz or 96 kHz (sometimes 192 kHz) for the sample rate, passing high-quality mixes to the mastering engineer, which are then dithered down to lower quality versions for release (CD and MP3). Please see the postproduction chapters (Chapter 12) about dithering and submitting mixes to aggregators.

Headphones ("cans") can be a popular way of working, but many professionals do not routinely use them. To many they can present a problem for a wide range of reasons, but also provide some large positives too. They can be considered a negative as headphone units can vary hugely and in times gone by most of the listening public would not be listening on them.

Monitoring on headphones can still be a positive inasmuch as the finer details can be easily heard (with the right cans of course). This latter point is the reason that so many still employ high quality monitoring headphones for this purpose.

The quality of the headphones can always be disputed whether that be the drivers themselves, or the headphone amplifier driving them. Either way, production teams seemingly prefer to use professional monitoring systems when working.

However, there is a compelling argument that we put forward in this chapter that much of the music-buying public consume their music using in-ear headphones on the move and as such we should monitor for this market! What are your views?

HEARING

Introduction

Our hearing system—our ears and how we process the information they provide—are undoubtedly the most precious tool of our trade. The ears are delicate devices and should be respected and utterly cherished. In this section we'll be looking at how the ear perceives sound in an irregular and nonlinear fashion and how this affects the way in which we work with and listen to music.

As you would expect, we'd be remiss by not touching on hearing loss a little. But before we can do so, we believe it important to delve a little into why we hear the things we do in the way we do, not necessarily in a scientific or medical way, but from a practical stance that allows us as music professionals to appreciate the variety of ways of listening and how they affect what music we produce.

Human hearing

Listening is dependent on a complex device, and even more complex to comprehend interpretation. The device is of course the ear. The interpretation is of course what we consider our auditory perception: the way in which our mind interprets the information it is provided from the ear.

These two elements, working hand in hand, continue to pose quite a few questions. Recently the medical profession has begun to understand how we hear and interpret sound a little more. Research into auditory perception has improved the codecs behind the data-reduction systems we're all familiar with. There appear to be around the corner, at the time of writing, some new codecs that might provide much-needed sonic improvement.

Let's take this part by part and look at how the ear works and what aspects we as producers would do well understanding while working in the studio. Understanding those reactions that the ear/listening system demonstrate when under certain conditions can improve your perception of how your music should sound, but also how it will be consumed.

Therefore, in this little section we'll be looking at the signal flow, if you like, between the sound creation, ear, and what we understand of how the brain interprets the sound we hear. This section is not intended to be exhaustive, but a guide. In Appendix F-1 at the end of this book you will find many books for further reading which we recommend.

As we'll discuss later, there are reasons why we're prone to turning the volume up and up and up as we groove to a favorite record. There are other reasons for wanting to turn it up in the first place. Before we can make a true assessment of the listening, which we will do later in this section, we ought to be confident in understanding the symptoms of our hearing system.

The ear

The ear is split into three important sections. It is important as music professionals to understand a little about these functions and how the ear operates under certain conditions.

The outer ear is the start of the story. The pinna is the fleshy exposed part of our ear. The next part is the concha which attaches the pinna to the ear canal. People refer to this collection of parts as the ear itself. This is only one, yet important, aspect of the ear.

Have you ever wondered why we can identify a sound from behind us, from even above us, or even below? Well, much of this is down to the pinna. This exposed receptor provides us with clarity to sound coming from the studio monitors exactly positioned to hit them while you're sitting in the hot spot. However, for sound coming from behind the head the pinna provides some dulling in the higher frequencies. This therefore provides our brain with one piece of information that tells us the direction from which a sound might be coming. Another indicator is the time difference as a sound reaches one ear before the other.

Next in the chain is the ear canal. The concha and the ear canal combined offer the ear a boost at high frequencies of roughly between 2 to 6 khz, depending on the person (see Equal Loudness Contours and Perception). A purpose of this is not so clear but it is plausible that this is to ensure maximum sensitivity to the frequencies of our human conversation.

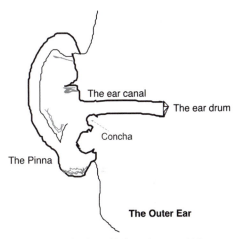

The ear's story begins with the outer ear, which contains the pinna, concha, and the ear canal providing two exceptionally important functions: funneling sound into the ear and directionality, among other things.

This proves an interesting point for us at this stage. The ear is a nonlinear device, being somewhat more sensitive to some frequencies than others. As such it is worth taking a small amount of time now to understand the effects of this and how this might impact on the music we record and produce, the first of which is explained in the sidebar on equal loudness contours and perception.

EQUAL LOUDNESS CONTOURS AND PERCEPTION

In the 1930s a couple of scientists engaged in some research attempting to appreciate how the human ear perceived frequencies against other frequencies across the spectrum. Their research culminated in the *equal loudness contours*.

(Continued)

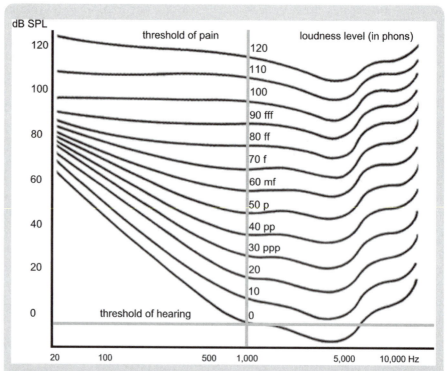

Fletcher and Munson's equal loudness contours precisely tell us why we need to be wary of the term *volume* as we work. The louder we turn up the volume the flatter the ear's response.

Fletcher and Munson set out to understand how much louder any certain frequency would need to be perceived as loud as the same perceived frequency at 1 kHz. The 1 kHz frequency is considered the center frequency between 20 Hz and 20 kHz—the human hearing bandwidth.

In equal loudness contours, the phon refers to a level at 1 kHz, thus in the diagram above, the 10 phon curve is the curve where all the frequencies have been perceived against 10 dB at 1 kHz. A 20 phon curve would be one measured against the perceived loudness of a 20 dB sound at 1 kHz and so on.

By testing a number of young people, Fletcher and Munson concluded that at varying different levels the amplification required to perceive frequencies at the extreme ends (low frequency and most high frequencies) could be large. For example, in the 20 phon curve (a 20 dB signal at 1 kHz) a gain of an additional 40 dB would need to be applied to a 30 Hz signal for the ear to perceive this at the same level.

This set of contours demonstrates two immediate things. Firstly that the ear appears to be most sensitive to frequencies in the 2–6 kHz range. Secondly that the ear's perception to sound is less linear than we might expect.

As you can see from the chart above, as the curves get louder, the perception of the listener becomes more equal, requiring less gain to frequencies at the low and high ends. This is of course of interest to the work we do because, as the listener turns up their tune, it will sound different to them.

We can never really automate these changes, only estimate their effect on what we hear. For example many hi-fi manufacturers have tried to overcome this feature. The first is the common "loudness" button found on most systems these days. Its purpose on introduction was to increase the bass and treble to assist those on lower volumes to perceive the bass and treble a little more, thus overcoming the Fletcher and Munson curves.

However, the increase in bass and increase in treble would remain static as the listener increased the volume on their favorite track, thus making louder versions very boomy and bright.

We remember Yamaha introduced a variable loudness control in the 1980s which would start from the max position (labeled "flat") and work backwards. As you pulled the control backwards not only would it attenuate the sound in the mids accordingly, but it would increase your perception of the Fletcher and Munson curves.

This was a neat idea rather than the on/off affair we are mostly left with today. This feature alongside bass boost in addition to the dreaded hall/pop/classical selections do nothing to necessarily improve the output of your music, but are regrettably important for music professionals to somewhat consider when working.

Wouldn't it be nice if everyone's hi-fi system was set to an optimum level so that music was perceived the same loudness in every setting? In Bob Katz's book *Mastering Audio* (Focal Press, 2007), he points out that this is the case in all cinemas as all the systems are set for THX, a surround sound protocol, and as such the level perceived in each movie theatre will be the same. Even Katz uses a set level in his control room for his mastering work. This is set to the 83 phon curve as it is where the ear is nearly the most flat. This is the same as the THX standard.

Again, how does this affect us in the studio? It is easy, as we've both been there, to get into the "vibe of the session," as Tim "Spag" Speight calls it, and to fall foul of wishing to crank up the volume on the control room monitors.

Every time we do this, our ears' perception has been altered and it could be argued that the brain needs to compensate a little. We're not suggesting that a standardized volume be used, as in the recording world this can be rather difficult to maintain, but an appreciation of the issues that raising the volume can produce is critical.

At the end of the ear canal is the tympanic membrane (more commonly known as the ear drum) sealing off the outside world to our sensitive middle and inner ears. The membrane is flexible and has the important function of acting as though it were the diaphragm on the most precious microphone of all. This receives the sound waves brought down the ear canal and, just like a microphone, converts the movement of the compressions and rarefactions into mechanical movements handled in the middle ear.

The middle ear connects the tympanic membrane to the inner ear, where all the processing takes place. One of its functions is to impedance match the incoming signal to the ear. Another is to help compensate for loud signals by acting as an attenuator.

This impedance matching is handled by the ossicles. These consist of a set of bones called the hammer, anvil, and stirrup. This clever little set of mechanics matches the rather free flowing sound in air that arrives at the tympanic membrane with the rather more resistant fluid within the inner ear.

Within the middle ear, there are a number of muscles which manage some ear protection. Their function is to resist the movement of the ossicles under loud sound pressure levels. In some ways you could consider this a limiter of one form or another.

THE ACOUSTIC REFLEX

The acoustic reflex is the effect in which the muscles within the middle ear contract while receiving loud sounds. A sort of limiter, if you like, for the protection of the fragile inner ear.

The acoustic reflex has another effect on the listening of a music lover. If the ear is exposed to loud sounds for a reasonable period of time, the acoustic reflex will kick in and the threshold of hearing for the ear is shifted higher. As such, more sound will be required to break through the instigated protection the acoustic reflex provides.

There is a funny link to this and how we listen to music. Many of us have been there where we've put on an album we really like and we've started to listen at a reasonable level. As track two kicks in and it starts to move you, you increase the level a little more. Track three arrives, and as the killer single off the album, you crank up the track. The fourth track starts powerfully and you creep up the volume a little bit more. By this point you may have been exposing your hearing via the hi-fi speakers or iPod headphones to high sound pressure levels for over 15 or 20 minutes.

It is at this sort of point that hearing damage can occur, but also it is critical that we understand how this affects the way in which we listen. Marry the Fletcher and Munson curves with how this has affected the acoustic reflex and threshold shift. It is clear that we want to turn things up to make the music *move* us, but the reason we want to do this is possibly because our ears are already trying to protect us. As we've already established, the louder we turn it up, the more frequency response of our hearing system alters. It is proposed that this might cloud the way in which we assess what we're recording or mixing.

The inner ear is where the main business of translating the sound arriving at our ear is turned into some kind of information that our brain can comprehend.

After sound has been converted into mechanical movement by the ossicles in the middle ear, this is then translated onto the oval window of the inner ear, known as the cochlea. The inner ear is enclosed and contains a fluid suspension all curled up in a snail-like structure.

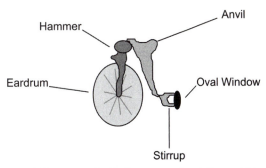

Hypothetically, opening this snail-like structure is like unrolling the Oscar's red carpet; we find something called the basilar membrane. The basilar membrane is the listening tool of the ear. Its function is to be the receptor of sound. "The Cochlea is the transducer proper, converting pressure variations in the fluid into nerve impulses." (Watkinson, 2004).

The middle ear contains the hammer, anvil and stirrup which together are termed the ossicles. Their functions are to transmit power from the ear canal to the inner ear while also protecting the hearing system from some damage.

Actually the basilar membrane contains a number of receptor hairs on it and on this set of hairs are a further set of hairs called stereocilia. These are the true receptors of the sound we hear.

The basilar membrane, if rolled out in a hypothetical linear fashion, is a frequency receptor that senses high-frequency sounds near the start of the cochlea working toward the end, which is low frequency.

So, along the length of the basilar membrane there are sections that receive certain frequencies. It is therefore important to note that if hearing loss occurs at one frequency, it is perhaps logical to conclude that this element of the membrane has been damaged.

As we have already established, the ear appears to listen to music in terms of small bands. The frequency response of the ear is split up into what are known as critical bands. These critical bands are considered the lowest common denominator of divisions the ear can discern apart easily.

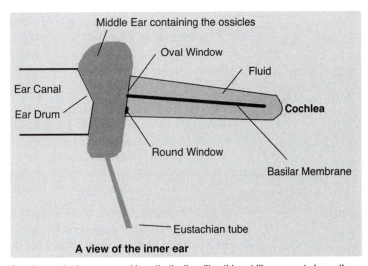

A view of the inner ear

Opening up the inner ear and hypothetically rolling this out like a carpet shows the basilar membrane and its connection to the middle ear.

Hearing loss

Hearing loss has always been with us, usually from the loss of upper frequencies as we grow older or in those unfortunate enough to have been exposed to loud mechanical devices (perhaps in factories) for long periods of time over the years.

Our current understanding of hearing loss alongside new, tight, health and safety legislation ensures workers are better educated and protected as much as possible from permanent hearing loss in the workplace.

For music professionals, the knowledge of hearing loss is better understood too. As such, many employ some rather excellent molded in-ear defenders in live music situations which attenuate frequencies as equally as possible, as opposed to the early stuff-them-in types which reduced high frequency.

For performers, the same technique has been employed in a way for in-ear monitoring, which no longer necessitates a set of powerful and loud wedge monitors at the front of the stage. The in-ear monitors can eradicate sound outside of the ear and allow a pristine musical mix in the in-ear monitors. In addition, they can be turned down (if desired) to reduce the exposure to high sound pressure levels that damage hearing.

There are a number of types of hearing loss on the ear mechanism, as described below.

Conductive hearing loss is essentially the middle ear's inability to move freely in the presence of thickened fluid. This is usually caused by infection or packed earwax. This usually can be repaired and medical attention should be sought.

Sensorineural hearing loss is more serious as this is where the inner ear is irreparably damaged and can no longer send the appropriate signals to the brain for certain frequencies. This type of hearing loss can be induced by too much loud exposure to sound but can also be induced by illness or physical damage to the head.

Mixed hearing loss is, as the name suggests, a mixture of the two previous types of loss. Meanwhile, unilateral hearing loss is where there is hearing loss to one ear. This can be a hereditary condition, or can be caused by external symptoms such as infection or damage.

The above hearing loss types produce effects on the ear and the perception of sound. In the music industry, noise induced hearing loss (NIHL) is probably the most common type of loss. Much research has taken place in recent times about the loss of hearing and its link to prolonged exposure to high sound pressure levels (SPLs). Listening to high SPLs for long durations of time should be avoided, as well as keeping the monitoring level down and taking regular rests from the control room.

HEARING LOSS TIME BOMB

Could there be a hearing loss time-bomb just waiting to happen? Essentially the relatively modern phenomenon of listening to music on the move via the Walkman in the '80s through to the iPod and other generic MP3 players has possibly given way to more intense and regular sound pressure levels being induced on our ears than ever before. Add to this the open backed construction of the typical headphones used in these situations. Thus, in order to hear the music more clearly, you simply turn up the volume to drown out the conversation on the bus or the train noise as it throws itself through the subway.

We wonder what effect this will have on the listening public later down the line. We propose a nonscientifically based set of rules (of thumb) about listening with open backed earbud style headphones. The first rule of thumb is that if someone sitting next to you can hear your music from your headphones then this is too loud for you. The second is that if you cannot hear the conversation around you in some semblance of clarity, then it's too loud!

WAYS OF LISTENING

Many people refer to something known as the *cocktail party effect*, a phenomenon where the listener finds themselves in a crowded room, perhaps a cocktail party, with significant background noise. All the listener's focus is naturally on the conversation with the person in front of them. This is an amazing ability to discern noise subjectively dependent on the focus of the conversation presented to the listener.

However, concentration on their topic can suddenly be broken by the sound of their own name or something that catches the mind's interest and the listener's head turns. This suggests to some extent or another that, although we're consciously concentrating on our own conversation with the person in front of us, subconsciously we are able to listen and, more critically, process some key aspects of the background noise enough to be alerted to key words that interest us.

These abilities are ones we as producers and engineers utilize a great deal, whether knowingly or not. There are many ways one can interpret this cocktail party effect described above. We can conclude, perhaps, that the listener's mind is always listening to everything whether they wish to or not, while being able to discern the most interesting conversation in front. Can parallels be drawn from this hearing ability to many who focus mostly on the vocal in their favorite pop song?

Can we also conclude that there is a lot of audio information (in the party and any piece of music) that is missed, or simply not appreciated or processed? Is this a shame? Perhaps in a musical sense it is, as the listener does not always process all the sounds coming toward them.

This is, of course, one premise studied and adopted in part for complex data reduction algorithms, allowing us to reduce full pulse code modulation audio (PCM data is the term often referred to as CD quality with no data compression) to smaller forms such as that used in data reduction systems. Data reduction systems in any significant technical detail are outside the scope of this book, but are of intrinsic importance to the engineer and producer. As such we will refer to data compressed forms of music and how producers can deliver on these formats.

Wouldn't it be nice if you could, as a music professional, switch between the two states of listening—listening to the person in front of you (the focus of the audio material) and listening to all incoming audio information—with almost equal appreciation?

The ability to switch between "ways of listening" can be very powerful in allowing access to vital information about the music you're working on for a variety of discerning buyers. If you can hear the material on many different levels, then you can assimilate how this material might be perceived by the background radio listener; the listener who sings along to the vocals with that as their focus; the instrumentalist who concentrates on their particular instrument in detail; and to fellow music professionals who, like you, analyze the material in many ways. Let's take a look at this in a little more detail.

Listening foci

> **CAVEAT**
>
> It is, of course, important to acknowledge that to a certain extent these different ways of listening are and can be interrelated and therefore are not clinically separated. But for the purpose of our discussion, we have made these slightly more exclusive in order to give some clarity to the subject.
>
> Above all, we realize that experience assists greatly in this area and this is something that cannot be taught. However, at the end of the book we intend to give signposts to other authors and texts which will help you improve your skills in this area. We introduce various ideas and concepts; however, for further detail and depth please refer to the Recommended Reading section in Appendix F-1.

In the introduction to this chapter, we already quickly identified some peculiarities in the way in which we listen and how our brain interprets the information it is provided.

There are so many ways we could analyze the way in which we listen to music or sound. We'd like to start with what Bob Katz, in his 2007 book *Mastering Audio, The Art and The Science* (Focal Press), refers to as passive and active listening. This is fairly straightforward to understand insomuch as passive listening is the background radio you may sing along to in the kitchen, and active is the concentrated focus on listening, whether that be a particular element or elements (or indeed the whole) of a certain piece of music.

This suggests the level of concentration or focus we're giving to the music. However, what if we wish to listen to certain things or a range of things or indeed the whole of the audio played to us? For the purposes of this book the change in focus between the minutiae and the song as one entity is what we refer to as *micro listening* and *macro listening*.

Macro listening refers to a higher level of listening. If we think in terms of altitude, this would be appreciating the whole of the town's landscape below over a large distance. This allows the listener to listen to the whole collage or sound (the song). This is not the same as passive listening, where the music might wash over you. This is focused listening, but on the whole of the content of the song as much as possible at the same time and not with any particular bias to elements. This is almost an element of detachment, which can offer you great objectivity when you need it.

Micro refers to the innate listening ability to be able to focus on singular aspects of the track (the conversation at hand when thinking back to the cocktail party effect). Using the altitude analogy, this would be zooming in on a particular street in the town landscape and maintaining focus on its surroundings.

The ability to listen actively and passively, micro versus macro, is incredibly valuable to all engineers and producers and can be honed. The ability to listen in detail to all information coming at you from the monitors at the same time is something that engineers spend years training themselves to do, but perhaps never quite achieve. Sometimes this has a negative effect inasmuch as it is very difficult to switch off. Tim "Spag" Speight, a former engineer for PWL, says "I can't just listen to music for enjoyment. I just can't do it". Speight goes on to suggest that he spent so much time analyzing music that it is too difficult to switch that inquisitive ear off.

However, there can be a downside in that some engineers have reported that they cannot employ the cocktail party effect in such high background noise environments. Research suggests that the loss of this effect is to do with hearing loss (something all music industry professionals should take time to research). How this affects sound engineers and music professionals is not so well-evidenced. Perhaps it is because they have trained their brains to listen in a macro way, listening to everything at the same time and not just following the tune or guitar solo? It is advised that engineers and music production professionals get a hearing test to be sure.

Holistic listening, as we refer to it, is listening in even broader terms than macro. Holistic listening could be likened to the listening process employed by a mastering engineer when working through a project. Here she may be concentrating with macro listening (in terms of each individual track) but will also be concerned with the holistic. In this sense the holistic listening refers to an appreciation of the whole album, or program of music, as though it were something smaller and tangible.

While a songwriter might compose a song over a three-minute period, or an artist might compile an album's worth of songs, a DJ meanwhile will mix together his selection of tunes, his set, over an hour (Brewster, 2000, *Last Night a DJ Saved My Life*). Within the set there will be peaks and troughs akin to the internal structure of the three-minute song, or the peaks and troughs of a well-compiled album. This is something we refer to as emotional architecture.

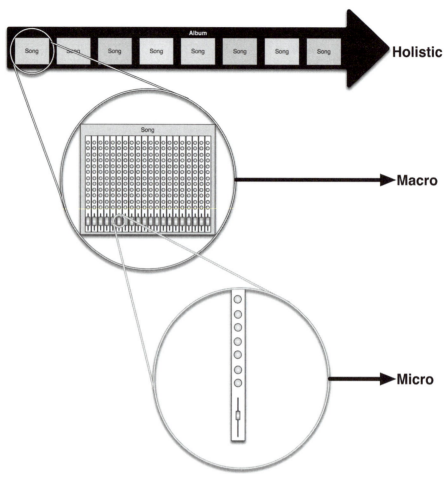

The different altitudes of listening can be holistic, or looking at a whole album, or program of music, all at once; macro, or considering the song at face value; or micro listening, which allows you to focus in on small aspects of the audio.

A mastering engineer works in this holistic manner by presiding over a whole album, gelling together the various tracks which may have been mixed by different mix engineers in different ways, ordering and organizing the material so that the flow is maintained.

Holistic listening skills should be developed by the producer in order to assess the overall sound of the project or album and therefore, in turn, realize the sonic aims of the project. Close links can be made with the *vision* aspect of the producer's role (see Chapter C-2, Pre-production), insomuch as the sound of an artist and his album could be foreseen and understood at the very start of a project.

Developing listening foci

Developing skills to listen in these ways will come naturally to many of us. We will appreciate that we can all focus in on one instrument in our favorite song. Focusing in on something we understand such as our own instrument, or the lead vocal we sing along to, is something that will be simple to do.

Developing skills whereby you can listen to a track and analyze many of its components at the same time, or appreciate the spread of energy in each frequency band, will only come with practice and plenty of active listening.

Practice makes perfect! So how should you go about this? The next section, "Listening Frameworks for Analysis," might be a good starting point.

To try to get to grips with listening, one must ensure that the listening environment is correct, with loudspeakers set up as suggested earlier in this chapter. The key is, as you will have no doubt gathered, to listen as much as possible. Not necessarily the passive listening we all engage in, but the eyes-shut active listening we've been speaking of. Closing one's eyes can improve the attention you can afford to the sound coming to your ears. Try this as a method to gain additional focus to the suggestions below.

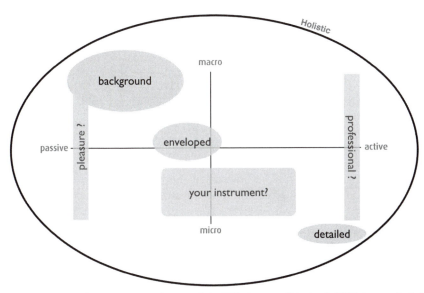

How do you listen? How might you choose to listen? What are you listening for? This image charts four quadrants of listening: passive and active against micro, macro and holistic. How would you plot these listening activities?

To develop skills in listening focus switching, just place a busy track which has large amounts of instrumentation on your accurately set up monitoring environment. Next, attempt to identify the instrumentation you can hear by switching from whatever your normal level of listening is and then focusing on one instrument; then zoom out to take in as much of the music as possible, perhaps even trying to get to holistic level where you're above the music, just assessing frequencies and intensities in aspects of the sound.

While this might be perceived as a skill predominantly for the engineer, it has uses for the producer also as it hones the listener's ability to be objective—a skill very much required for the job of producer.

Specifics

So far we've been mainly concerned with altitudes of listening, or ways of focusing in on what is important in a track. Focus into the relevant zoom level is one thing, but what do you do when you get there? What is it you're listening for? What are the parameters by which you'll analyze and use what you hear?

With micro listening we could expect to solo in on an instrument on the mixing console. How do we assess it when we get there? Do we look at the engineering: its sound, its timbre, its frequency range, its tuning, the capture, or the performance and all that entails?

At this level of listening we'll also be concerned with the minutiae of the performance in terms of tuning, timing, and delivery. We can be very analytical when a track is soloed, concentrating on every detailed aspect of the sound, pass after pass. It is easy to enter this level of analysis when the instrument is presented to you in this manner.

How we approach separating out the same instrument when presented to us immersed in a whole track is a skill that can be developed. Again, to some this will come naturally, while in others this will have to be developed. Simply focus in on one aspect (best to try an instrument you know well, or the vocal) in a busy track and try to listen to it in isolation, attenuating the other aspects from your listening experience. This can be hard to do, as it is one signal intertwined together with perhaps another 23 tracks of audio information. However, the ability to focus and discern what that track's worth of audio is doing, and thus its effect on the mix as a whole, will become an invaluable skill. Personal-solo, if you like.

However, once we've separated out the audio or instrument we wish to focus on, we need to concentrate on making an assessment of it both sonically and musically. As a producer, your role will vary from the person who says that the sound being produced is not the "right" sound for the style you're wishing to convey through to the performance is not quite as you'd expected, and so on. The call is potentially yours to make. Therefore, it is important that you hear and appreciate what is happening with that instrument.

In a previous book (*From Demo to Delivery: The Process of Production*) we proposed the employment of a self-developed *listening analysis framework* which dissects the music into a number of headings, acting as prompts for the listener to make an educated assessment of certain parameters. We'd like to develop this in the next section by looking into each section a little further.

LISTENING FRAMEWORKS FOR ANALYSIS

Developing a listening framework for analysis is a way in which analytical listening in music production can be taught. This short section is inspired by the excellent work of William Moylan. Moylan is one example of best practice in this area. We wish to convey one method we have used to kick-start many of our students over the years into track analyses.

Example listening framework

Frameworks for analysis are the starting points to dissect how a recording has been put together objectively and how its result actually sounds. Below is an example of what we might consider the bare essential elements and a structure by which listening can be analyzed. This framework will need to develop and mature depending on its use and function.

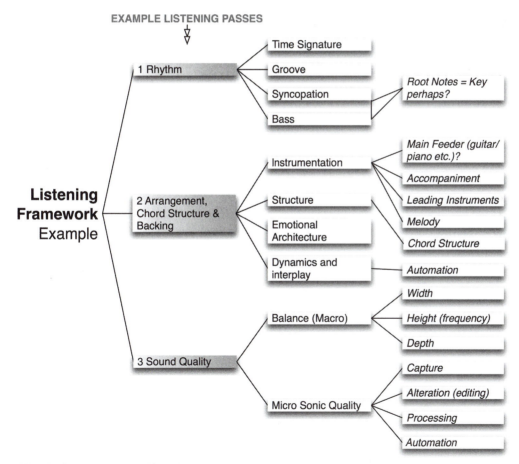

A listening framework such as this basic example can provide a checklist against which overall sonic and production qualities can be measured.

Listening with a critical ear is a personal pursuit and will vary widely from person to person. In the framework example above, we have introduced the idea of using listening passes (passes refer to the number of times the tape passes the playback head or the number of times we listen to the piece in the DAW-led world we now live in). In this example we listen three times through and break these main areas down yet further to assess them and comment on them. It should be mentioned that more passes are often required.

This branch diagram could be expanded to offer exacting questions from each of the prompts here if required. With more experience one or two passes should suffice for the simple list shown above.

PASS 1

In this example, pass 1 is entitled *rhythm*. The first thing that can be assessed is the form of the drums, or rhythmical elements. Analyzing this part of the recording can reveal the time signature and offers some insight into the groove of the piece. Taking this one step further, linking the bass of the song within this pass, we can learn much about how the song is constructed, whether there is syncopation, and what the bass is playing. This pass could of course be expanded to the types of rhythms and how they interact with other instrumentation.

PASS 2

Pass 2 turns attention to the more musical aspects of the recording. In this pass we're concerned with the arrangement, chord structures and backing—essentially the way in which the music is constructed. This includes an analysis of the instrumentation. It is here that an identification of the main feeder can be made. The main feeder in a song is the instrument that drives the song. Not a vocal hook necessarily, but an element that makes the song memorable and often the main driver in the backing. It may be the catchy riff, or the solid pounding of piano chords. We look here also at the accompaniment: what makes the whole composition tick behind the scenes sonically? Finally the melody can be assessed.

The arrangement section is broken down into three elements: structure; emotional architecture; and dynamics and interplay.

The structure of the piece, if it is your own composition, is already known but as new tracks are introduced and these skills develop, structure will be an informative place to gather plenty of information of the construction of the song. It is at this stage we can identify if there are things that can be changed and manipulated.

We next assess the music's emotional architecture. Emotional architecture is the way in which the music builds and drops. This is not an assessment of dynamics (covered next), but an assessment of how the music affects the listener through its intensity. This is something that can be drawn graphically as shown below.

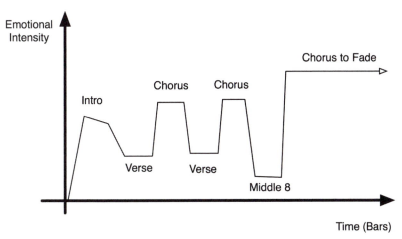

Emotional architecture graph showing the intensity and power of a track as it delivers its choruses and a large ensemble at the end.

In this graph, the emotional intensity, or power of the music, can be shown and tracked. Assessment of many popular tracks can often show themes or perhaps new ideas of how to achieve this variety in emotional intensity.

Next we would quickly assess the dynamics of the instrument and how this interplays with the track as a whole; for example, the dynamics of a given instrument and how it might be integrated into the mix using automation, or other mixing techniques covered later in this book.

PASS 3

On the last pass, we assess the sound quality of the music and dissect its construction. This is split into two areas and here we introduce the terms *macro* and *micro*. In this case we refer to macro, meaning the balance of the instruments together, and micro is the focus on individual elements within the mix and their sonic quality.

Here we refer to macro listening as broad listening subsections: width, height and depth." The first refers to stereo width and the use of panning. We would wish to make an assessment of how the stereo field has been made and constructed. Height, in this instance, refers to the use of the frequency range, or spectrum. Some artists make full use of all the frequency ranges with deep, but focused, bass and excellent detail at the top end, while other pieces choose to limit the overall bandwidth they occupy. It must be noted here that height can also refer to the vertical presentation of a sound. Last is depth, which is an assessment of the presence of any given instrument or the mix as a whole. Is the mix deep, presenting instruments that appear to come from behind others, or are all instruments brought up to the front in the mix?

As previously introduced, micro listening is an assessment of the internal elements that make up the sounds in our mix. How instruments have been captured and whether there are issues inherent in their sound can be gauged. Listening in a focused manner using a good reproduction system can also reveal edits in material that were not supposed to be heard: drums that were replaced with samples, and vocals that were automatically tuned. Estimates of the types of processing that were used in achieving the mix can be made.

Automation now plays a huge part in mixing and presenting the micro elements. Can an assessment of the automation moves be assessed on very close inspection? How does this shape the presentation of the music?

Framework analysis

Once an overview of a framework is decided on, we advise that comments be scribbled down in response to each of the areas in the framework. A personalized proforma could be created. An example has been given below, based on the basic framework given. Any proforma document should be well-designed in such a fashion that it takes little time to become accustomed to, allowing reliable information to be recorded about the song in question. How this is notated will be a personal method based on whatever can be notated in the boxes.

In time, this skill should not require a lengthy written document or framework proforma as suggested here, but something to jot simple notes down of pertinent points observed. This framework acts as a checklist that one day should become second nature.

In the version below, a rough analysis is made using the 1978 Genesis song "Follow You, Follow Me." The analyst has chosen to write comments that might be pertinent to recording the track's qualities. For example, a list of things to listen out for in the next pass, such as the key, which could be in some dispute. These have been adorned with question marks and can be reference points for later investigation on additional passes. Other listeners might simply wish to rate the sound and use a star rating, although this would not give exacting points for discussion or further analysis.

As discussed, this example, or method is far from exhaustive and should be expanded and personalized where possible to suit the genre and purpose of the analysis. Analysis is ongoing research for all of us and only by listening and questioning will we learn more about our craft.

We have provided another four listening analyses for you to look at. Try and listen to these tracks and see if you concur with the analysts views. Let's listen!

'FOLLOW YOU FOLLOW ME' GENESIS 1978.

Listening Framework Example   Rating/Comments

1 Rhythm

- ☐ **Time Signature** 4/4 ?

- ☐ **Groove** Fairly Straight. Meandering.

- ☐ **Syncopation** Gentle integration between instruments.

- ☐ **Bass** → Guitar. Locked fairly to drums. Good Interplay.
  - ☐ Root Notes = Key perhaps? Percussion

2 Arrangement, Chord Structure & Backing

- ☐ **instrumentation** Electric Guitar, Bass, Synths, Drums, Vocals, Lead Guitar, BV's + Percussion.
  - ☐ Main Feeder (guitar/piano etc.)?
  - ☐ Accompaniment Guitar riff (muted) + Mono lead Synth Organ and Bass.

- ☐ **Structure**
  verse/chorus/middle 8/chorus etc.
  INTRO / VERSE / CHORUS / VERSE (2) / CHORUS / SOLO (Keys) / CHORUS / CHORUS Repeat to fade.
- ☐ **Emotional Architecture** Fairly Straightforward →

- ☐ **Dynamics and interplay**
  dynamics between instruments that bring out different flavours in the arrangement and thus interplay between sounds Fairly Static Throughout.
  Some play between bass, guitar + drums
  - ☐ automation ? Perhaps. Gently balanced which works well.

3 Sound Quality

- ☐ **Balance (Macro)** Solid balance between instruments. Vocals weak and far back in mix
  - ☐ Width
  - ☐ Height Good use of delayed effect + phase on
  - ☐ Depth Guitar riff. Good use of width.

  Pretty two dimensional Frequency range Static but fairly broad for the music
- ☐ **Micro Sonic Quality**

  - ☐ capture Solid capture. Good drum sound, if a little low in the mix
  - ☐ alteration (editing) Difficult to assess, Given the track's age it is unlikely
  - ☐ processing Fx includes delay, phase, reverbs & think
  - ☐ automation again difficult to assess.

A fictional example of a completed listening framework marksheet.

Listening Framework Example - Alanis Morrisette (Producer Glen Ballard) - Hand In My Pocket

1   Rhythm

- ☑ **Time Signature**
  Straight 4/4 approx. 110bpm

- ☑ **Groove**
  Supplied mainly by drum loop

- ☑ **Syncopation**
  Syncopated parts drum loop, guitar and bass interaction.

- ☑ **Bass**
  Basic bass line entry in 1st chorus section, slightly more developed part from 2nd verse onwards.

  - ☑ Root Notes = G/C/D Gmajor Basic synth bass line playing chords V,I in verses.

2   Arrangement, Chord Structure & Backing

- ☑ **instrumentation**

  - ☑ Main Feeder (guitar/piano etc.)? - electric guitar (looped?)
  - ☑ Accompaniment - Drum loop/s, Synth Bass, Elec guitar, Organ, Harmonica

- ☑ **Structure**
  INTRO, VERSE, CHORUS, VERSE, CHORUS, INSTRUMENTAL (Harmonica entry), CHORUS, VERSE, CHORUS, OUTRO.

- ☑ **Emotional Architecture**
  Generic layering to increase/decrease. Build from sparse intro section into 1st Chorus, 2nd verse more energy into 2nd Chorus, drop for 3rd verse into bigger more energy final chorus/outro section.

- ☐ **Dynamics and interplay**
  dynamics between instruments that bring out different flavours in the arrangement and thus interplay between sounds

  Dynamic of new snare/drum sounds for 2nd verse/chorus. Big dynamic drop for verse 3 with only original drum loop/guitar exposed with basic bas line. Basic interplay between accompanying instruments.

  - ☐ automation

3   Sound Quality

- ☑ **Balance (Macro)**
  Solid balance with prominent drum loop providing groove. Vocal sits relatively high in the mix, with guitars and accompanying instruments underneath.

  - ☑ Width - Good use of width with doubled tracked guitar part panned hard left and right, stereo keys/organ adds width, with other layered parts sitting in between. Harmonica to left with Bass & vocals centre.
  - ☑ Height

    Frequency range fairly static other than use of 'filtered' drum loop and full bandwidth drum loop. Later introduction of bass adds a lower layer.
  - ☑ Depth

    Relatively two dimensional, limited depth added with use of reverb on vocal and harmonica.

- ☑ **Micro Sonic Quality**

  - ☑ capture - Mostly sequenced instrumentation, good vocal capture, good harmonica capture.
  - ☑ alteration (editing) - Vocal editing very noticable in 1st verse, detectable drum loop start/finish.
  - ☑ processing - Compression, gating(?) Reverb, Vocal reverb introduced on 1st chorus, filtering, chorus effect on bass?
  - ☑ automation - Difficult to assess

Listening Framework Example - Amy Winehouse (Producer Mark Ronson) - Love Is A Losing Game

1 Rhythm

- ☑ **Time Signature**
  4/4 approx 100bpm

- ☑ **Groove**
  Relaxed groove, laid back slow late 60's soul/RnB style.

- ☑ **Syncopation**
  Syncopation provided by drum and guitar 'stabs' part.

- ☑ **Bass**
  No real 'bass' part.

  - ☑ Root Notes = Key perhaps? C,F,G - C Major

2 Arrangement, Chord Structure & Backing

- ☑ **instrumentation**
  Vocals, Guitars, Strings/brass, Drums/percussion, Vibes

  - ☑ Main Feeder (guitar/piano etc.)? Guitar (inc. stabbed part) ,Strings
  - ☑ Accompaniment - Drums, Vibes

- ☑ **Structure**
  Verse1, refrain, Verse2, refrain, Verse 3, refrain.

- ☑ **Emotional Architecture**
  Relatively linear, some build with srting section. no real contrasts

- ☑ **Dynamics and interplay**
  dynamics between instruments that bring out different flavours in the arrangement and thus interplay between sounds

  Dynamic 'swells' in the string arrangement give some shape/ addition and subtraction of guitar and low level brass also give some dynamic constrast. Faiirly constant with only subtle changes overall.

  - ☑ automation Difficult to assess although style/recording possibly negates needs for automation

3 Sound Quality

- ☑ **Balance (Macro)**
  'Retrospective' style of mix, with Spector-like arrangement with main focus on vocal. Guitars/string section resonably prominent in mix, with other brass/vibes at low level.

  - ☑ Width - stylelised 60's Spector-like positioning with hard left/right panning. String section to right. Vibes to right ,Main guitar left with guitar 'stabs' right.
  - ☑ Height - Frequency range static with no dramatic chnages throughout, strings do give sense of bass depth in places
  - ☑ Depth - Depth created by used of various reverbs and 'natural' ambience to capture. Typical 'large' 60's style sound.

- ☑ **Micro Sonic Quality**

  - ☑ capture - traditional capture, with spill and ambience used. Acoustic instruments only - No sequenced parts
  - ☑ alteration (editing) - no evidence to suggest this - although possible vocal comp'ing
  - ☑ processing - effects used include, spring, plate and hall reverbs, some compression on vocal (and possibly other parts?)
  - ☑ automation - unlikely and not apparent.

Listening Framework Example - Michael Jackson (producer Quincy Jones) Workin'
Day and Night

Rating/Commen

1 Rhythm

- ☑ **Time Signature**
  4/4 approx 120bpm

- ☑ **Groove**
  Pop/funk groove created by 'swung' percussion and bass/piano lines

- ☑ **Syncopation**
  Heavy syncopation provided by interaction between drums,percussion,
  bass line, EP,

- ☑ **Bass**
  Repeated bass line/phrase over two bars

  - ☑ Root Notes = Key perhaps? E Major?

2 Arrangement, Chord Structure & Backing

- ☑ **instrumentation**

  - ☑ Main Feeder (guitar/piano etc.)? - Piano/Bass synth
  - ☑ Accompaniment - Drums, hand percussion (shaker/triangle/agogo bell/hand claps), roto toms,palm muted guitar, guitar, Bass Guitar (slapped) Brass (Trumpet,Sax Alto/Tenor) BV's.

- ☑ **Structure**
  Intro/Verse1/Pre Chorus/Chorus/Verse 2/Pre Chorus/Chorus/Instrumental (Brass break)/Chorus instrumental/Verse 3/Pre Chorus/Chorus/ instrumental 2/Chorus vocals to fade.

- ☑ **Dynamics and interplay**
  dynamics between instruments that bring out different flavours in the arrangement and thus interplay between sounds

  Overall dynamics relatively static with fluctuations created by layering of parts. Much interplay between percussive sounds

  - ☑ Emotional Architecture - High energy groove throughout, with only slight drops during instrumental sections

3 Sound Quality

- ☑ **Balance (Macro)**
  Solid balance with strong emphasis on bass line and percussive groove. Punchy multitracked vocals and brass section. 'Tight' overall sound quality.
  - ☑ Width - Good use of stereo width with percussive elements panned hard left/right, also guitar parts left&right. Brass section multiple tracked and spread across the image, same with BV's.
  - ☑ Height - Good frequency bands with top end enhanced via percussion (shakers/triangle) perceived height gained via top brass and some hi register vocal ad-libs.
  - ☑ Depth - Varied use of reverbs adding depth to brass and vocal parts in particluar.

- ☑ **Micro Sonic Quality .**

  - ☑ capture - Mutlitracked vocal parts and brass section, overdubbed hand percussion, live instrumentation with some synth parts played live.
  - ☑ alteration (editing) - Possible editing on analogue format but difficult to assess from recording
  - ☑ processing - Compression/EQ of percussive parts, brass and vocals. Revebs used across vocals, brass. percussion, guitars
  - ☑ automation - Difficult to assess. but some possible mix automation

Listening Framework Example - The Beatles (Producer George Martin) - Hey Jude

1 Rhythm

- ☑ **Time Signature**
  4/4 approx 100bpm

- ☑ **Groove**
  Straight slow rock/pop feel

- ☑ **Syncopation**
  Standard syncopation between drums & percussion and piano.

- ☑ **Bass**
  Bass Guitar, & left hand bass piano

  - ☑ Root Notes = Key perhaps? F, C, Bflat, - F Major

2 Arrangement, Chord Structure & Backing

- ☑ **instrumentation**
  Vocals, Piano, Drums, Percussion, Bass guitar, Acoustic guitar, BV's, Strings/ Brass,

  - ☑ Main Feeder (guitar/piano etc.)? Piano
  - ☑ Accompaniment - Drums, Percussion, Bass guitar, Acoustic guitar, BV's, Strings/Brass,

- ☑ **Structure**
  Repeated verse sections, with refrain, Instrumnetal section with main vocal refrain/melody repeated to end.

- ☑ **Emotional Architecture**
  Smaller and more sparse intro and first sections with slow build and layering. Larger 'climatic' 'nstrumental /gang vocals mid section repeated to end with fadeout

- ☑ **Dynamics and interplay**
  dynamics between instruments that bring out different flavours in the arrangement and thus interplay between sounds

  Some dynamic contrasts particularly between vocal parts,some interplay between vocal and instrumental parts.

  - ☑ automation - no automation due to technical limitations of the equipment at time of recording

3 Sound Quality

- ☑ **Balance (Macro)**
  *S*olid balance between instruments with focus on main vocal/piano at start. Balance shifts with BV's and gang vocals in mid section to end.
  - ☑ Width - Wide use of 'stereo' image typiclal 60's extreme panning with Drums left, Piano right, lead vocal leff, Acoutic guitar hard left, percusion centre/left.
  - ☑ Height - Fairly static throughout although use of tamborine and percussion gives some perceived height.
  - ☑ Depth - String/brass section gives some depth

- ☑ **Micro Sonic Quality**

  - ☑ capture - Live capture - acoustic instrumentation.
  - ☑ alteration (editing) - limited only bouncing down due to track limitation
  - ☑ processing - reverb/echo chamber
  - ☑ automation - no automation due to technical limitations of the equipment at time of recording

THE PRODUCER'S WATERMARK

As we've discussed, each producer will have a watermark or range of watermarks for types of material. It is important to note that the watermark can work in two ways. It can be a statement, or sonic identity, of a particular producer's style. Or it can be an application of certain tools for a desired outcome.

Some producers will have a distinct, intrinsic, natural style. It could be argued that Phil Spector is one of these. His wall of sound became one of the most notable watermarks in production history. Not only is it permanently linked with Spector, it is also simply the most famous producer's watermark, often imitated. Not every producer has the benefit (or disadvantage) of such a distinctive watermark. Too distinctive a watermark can equate to pigeonholing, which might lead to being branded for certain limited styles of music or artist. Having no watermark can leave the producer indistinct and unable to stand out from the crowd.

We talk in this book about watermarks being a sonic musical outcome; however, a watermark can equally be a reflection of your professional reputation. Long-standing successful producers such as Trevor Horn continue to enjoy great success producing a variety of styles of music. He is seen within the industry as not only a pioneer of music production but also as a "safe" pair of hands to insure a relative amount of success from a project. Horn has continued to develop a reliable team of professionals at the top of their game alongside his wife Jill Sinclair to develop contributory projects such as ZTT, Sarm Studios, which all feed back to support his production work. We also talk in this book about being a business, as this can become equally important as the toolkit that defines your sonic style discussed in this chapter.

SECTION B
Being It

Being a music professional is quite different to being an accountant or some other trade. Our work days will usually involve being in front of a set of monitors (both the loudspeaker and screen variety these days!) listening, composing, recording, mixing, or mastering.

This is what we do. However, most people that relax at a pub/bar after work will listen to music in the background. Then they may amble off home to watch a DVD with a film score included.

Music is all around us. Therefore our work is all around us! Do we ever get a break? Actually most of us will not mind this intrusion, but it does make us music professionals different. We're always on duty, we're always listening, we're always analyzing. To some extent we're *always* working.

This chapter introduces you to the skills and attributes that a producer has. By no means is this a job specification for some imaginary post, but a peep at the lifestyle and activities undertaken and important areas for attention as you build your career.

COMMUNICATION

Producers need to be fantastic communicators who have depth and consideration for their artists' views. It will not always be possible to pander to them, as producers have to take executive decisions. Gaining trust from the artist will allow for any necessary experimentation. The ability to listen, and listen with empathy and in detail, will ensure better understanding.

If we were to read that mystical job description for the role of the producer, part of the list would no doubt request honed listening skills, good communication, diplomacy, charm, modesty, and humor, in addition to being a good politician as required.

Producers have to manage their relationships with their artists on many different levels. The music recording studio can be a close environment within which to work for many a long session. Being someone who can make the session

What is Music Production. DOI: 10.1016/B978-0-240-81126-0.00003-2

flow using a blend of skills is important. You will need to be, in part, a friend, a colleague, a counselor, and in some extreme cases a psychiatrist.

Consider the situation where a musician is bearing their soul in a song about lost love, death, or something deeply personal. This may leave them feeling exposed and perhaps less confident, or emotional as a result. You are the pilot of the ship known as the recording session, and keeping the artist in a productive frame of mind is essential.

An example of this is Quincy Jones speaking of a time in the recording of *Off The Wall* where Michael Jackson, while singing "She's Out of My Life" was actually crying, partly because of what he felt, and partly to deliver the tone of the song. Managing and funneling this emotion into the song is very important. But as a producer, how do you move on from this? Do you abandon the session for the day? Do you mix the track? Or do you move to another emotionally charged track on which the lead vocals need recording?

Your decisions and the way in which you negotiate them through communication will be the leading light required for the furtherance of the project. The style of communication will also be key to getting along with the artist. Some artists of course have a certain style about them and perhaps a certain genre-specific language which you will need to be aware of. You may need to pitch your communication to each different audience in a slightly different way. Gaining a sense of the vocabulary used in the studio, and in the genre, will assist the translation of ideas between the production team and the artist.

Given the emotional flow of an artist, they may choose to discuss the background to their music and their life. This can be important information, however tedious at first it might seem. Getting to know your artist will provide you with valuable information about what the song or album means and how it might shape up.

It's all part of being a good listener, which is an essential skill as a producer. Tommy D before a songwriting session with an artist takes this one stage further, to be prepared. He takes time to learn about what they have been doing and "where they're at in their life." "I always try to have a couple of [musical] ideas together before they come in. I'll go online and read about what they've been up to, and I look at their Twitter feeds and Facebook updates and see what they've been doing and where they're at." This allows Tommy to begin to engage with his artists from the get-go and perhaps to consider lyrics and music that might resonate with them.

In a similar vein to genre-specific language, the art of communication is not simply about what you say and how you say it, but a number of other factors too. We have been at pains throughout this book to illustrate the role of the producer as someone who guides the artist and the music through the production process. The producer is a breed of person that is unique and highly skilled in many a communication method concurrently. One of these skills is communication and another is how they handle themselves.

COMPOSURE

BODY LANGUAGE

It's not all simply about what you say, or even how you say it. It's also about how you are. Body language can be incredibly important in getting information across. As a producer, not only do you need to be measured in the body language you provide, ensuring it sends positive messages, but you also need to be alert to the body language offered from the people around you.

Understanding the body language of others can determine what they are really trying to say and from this you may be able to develop a plan of action that honestly fits, as the verbal communication may not stack up. You can imagine the scenario where a band is in the studio and the singer (the writer of the material) does not really like the way in which the guitar part is being turned into a rock guitar solo from the '80s. While the singer's words might be "sure, let Johnny play what he wants," the eyes and general slouched shoulders might tell another story.

How you attempt to alleviate the situation might be various. You may choose to use diplomacy and negotiate a simpler guitar solo in keeping with the indie rock of recent years, or to use multiple guitar tracks, harmonically arranged to produce a retro Brian May from Queen style solo, adding texture to the song, rather than a ripping '80s rock solo.

By the same token, you will need to temper your body language accordingly. There will be those album tracks you're not so keen on and you may not be able to say so for various reasons; perhaps it's the only track the bassist has written and hence has to have a place on the album. Anyway, you may choose to appear as keen and professional toward this track as the three No.1 singles about to fly out the studio.

ATTIRE

Producers can often be flamboyant individuals who have a style of their own. For example, P Diddy has a 'look' as does Mark Ronson these days. They have a style that looks confident and is just as much their image as the music. The same can be said for Dr. Dre and Timbaland in the urban genres.

The way you dress may be to impress and elevate your status, or it simply might be the way you are. Many producers and music professionals straddle a line of comfort (as you're in the studio for long periods of time) or style. We're not suggesting that you have to dress to impress, but you may choose to develop a style that is yours and that may convey the musical image you're after. We'll leave that up to you.

CONFIDENCE

The interviews and discussions we've enjoyed in the research for this book point toward one overriding common trait: confidence.

What is confidence? A good question to begin this section, we think. Producers are a funny breed insomuch as they are leaders, not in the military sense of the word, but in terms of respect. They do not have the ability in most sessions to dictate the music where they want it to go. Naturally there are exceptions to this, but on the whole, the average music producer is simply confident and an excellent interpreter and communicator of ideas.

In order to succeed in the industry, it is necessary for a producer to be confident in their decisions and outcomes. It is not always that easy, but many artists will be looking to the producer for confidence, a strong guiding hand through the process. If you like, an insightful interpreter.

In this chapter we'll take a look at the confidence that a producer typically possesses and how it manifests itself, how it can be increased and how to manage this development. Some do indeed say confidence is something you are born with, but it can be developed and enhanced for display, whether you believe it or not! Producing is about modesty, humility, creativity, and, of course, confidence.

What is confidence?

Confidence can manifest itself in many different ways. Many of the producers and music professionals we have met over the years are some of the most modest and lovely people you could hope to meet. They are good at listening and very good at suggesting their views on music and its direction.

Producers are good listeners, as their job is to interpret artists' ideas and help them realize them on a record. Their ability to suggest is not driven by an arrogance, but usually experience. There is an air of confidence that suggests they have a way of doing things which might, or at least should, work.

INNER CONFIDENCE

External confidence is the ideal outcome: gaining trust and then exuding an air of confidence in the session. This usually can only be created honestly from internal confidence. Internal confidence is something of a different ball game altogether and something that the self-help book industry makes millions from each year.

Confidence is not something you wake up with one day, and overcoming a lack of confidence can take considerable effort.

The reality is that the producer requires some form of inner confidence which is based on skill firstly, whether that is musical, engineering or a combination of the two, or in some cases business acumen. These are all fairly core skills in the studio today.

To develop this confidence comes from experience, many suggest. This can be experience in turning that skill and knowledge into practice, creating new and exciting outcomes for the artist. Or the experience of working with a range of artists and bringing their visions to fruition on a multitude of projects.

Having inner confidence is essential as this will provide external confidence, leading to an ability to empathize, be diplomatic and reactive to the emotionally charged situations in which you'll find yourself in the studio.

Being confident in your own abilities as a producer is a huge necessity in the studio. Being able to acknowledge your weaknesses and use them to your advantage in the session can help. For example, it can be helpful to collaborate with a partner as part of a production team. Finding a collaboration which works for you can assist in ensuring you're able to deliver aspects of the session, the album which alone you'd not be able to achieve. For example, a fantastic programmer to work with your fantastic orchestration skills, or songwriter to work with your skills in managing the session.

For example, you will collect a list of preferred musicians and engineers you choose to pool from; something we cover in Chapter B-2, Your People. Having a team that you trust implicitly will improve your confidence and ability to deliver on the recording session and ultimately the production. The artist may have come to you to produce her album, but it is your team behind you that is also called on.

EXTERNAL CONFIDENCE

Confidence being displayed and hopefully engendered in your musicians will come from you making good and insightful decisions based on their needs and ultimately delivering the goods. It also comes from solid communication as previously discussed. Being able to listen, use diplomacy, and steer the process to a final record will be managed by an air of authority in the studio.

If they, as artists, believe in you and the guidance you provide, then they will also have trust in the vision for the music and where it is scheduled to end up. However, external confidence is not simply about how you deal with musicians, but also how you conduct yourself in business too.

Music and musicians are forgiving; they know what they think they want, but because they spend much time perfecting their skill and their parts, they can lose a little focus on the whole, and gain considerable benefit from the guiding hand of the producer. However, they're placing strong trust in the producer to take their music and make it better, more what they envisaged, and make it sell (a tall order).

Working with a producer can be a learning curve for many musicians. Bands can be very precious about their music and what they have developed. It is, after all, their art and their craft. Musicians might find it difficult to manage to change their paradigm to involve and embrace an external person to the production process with new and seemingly conflicting ideas.

This is where the diplomacy and confidence that a producer displays can be beneficial. The confidence a producer exudes can allay fears within musicians and allow for ideas to be explored, where they might not have been accepted in times previously. The music needs your guidance and the artist must believe in you and your confidence.

ATTITUDE

It's all about attitude. Many producers we've interviewed have said that they expect their colleagues (engineers and assistants) to have a "can do" attitude, being responsive and one step ahead of the game. They must be able to anticipate what's about to happen by the series of cues they pick up on from the artist or fellow production team member, to make the process go smoother.

As a producer, you will no doubt exude the same can do attitude, listening to your artists' ideas and views and trying to calculate a way of making it happen. However, should you not agree with a course of action or decision, you will need to use your aforementioned skills in communication to discuss and diplomatically agree on a solution. There's always one good way of sorting out any arguments, time permitting. In these days of nondestructive editing, it's often a good idea to try out even the ideas your artist insists will work. If you think they will not work and are proved right, at least you've given the artist a good, honest hearing.

Having the right attitude, like trust, can come from experience and a desire to be productive in the studio environment. Your attitude should reflect that you are passionate about music, the art of music production, the artist, and the music they're creating. A positive attitude will make the session go well, but will also be fun! Remember they're long sessions; fun will make the whole thing a lot more enjoyable.

NETWORKING
It's all about who you know!

It's not always what you know, but who you know! The music business is a fickle business and is based on reputation. As old-fashioned as it sounds, who you know is still important. How do you get to know "who you know" in the first place? How do you get known and considered when the work comes up? It's simple: through a range of activity, of which one of the most important is networking.

Let's expand on this a little. It is not quite as sordid and opportunistic as it sounds. Thinking about it deeply, one can deduce that people who like each other work together. If you do not like the lead singer in your band, it is probably not always worth going on a long tour of America with him, as the tour bus could become a cold and lonely mobile prison. Even the best of friends in this goldfish bowl-like mobile home have arguments and times where things do not gel.

It is no surprise therefore that people that get on often choose to work with each other. It helps your friend out and it also makes the whole process more fun. This is not only the case in the music industry, but we'd dare say most industries. It boils down to trust. If you trust and rely on someone, surely you're more likely to give the work to them?

Of course, there are natural exceptions to this rule in that you may "put up with someone" because they are simply an amazing player and you can manage to endure a three-hour session with them, or longer if they're that good. There is a flip side to this, in that many people might choose to work with friends first, attributes second. "Oh, wouldn't it be cool to have Johnny in the band?" However, Johnny, while fun to be with in a bar/pub and a tremendous attractor to the female fan base, is not that great as a drummer. And being the drummer of course means that the band cohesion in performance and recording suffers.

The same can to a lesser extent apply to the producer. Many people will work with a producer because they like the previous album he produced with their closest rival, or they've heard amazing things generally. However, the obvious producer is not always the right producer for your project. Many producers work with artists over many, many years and provide cohesion.

What we're really getting at is that this is a people business. It is one of those strange industries that relies on the creative and the technical combined. Creative people are to some extent social animals generally and can excel in the networking arena. They may naturally shy away from the technical, although in the past 20 years there are breeds of both people who have adopted technology wholeheartedly to create quality music across all genres.

Why network?

If it's a people business, you'll need to get to know more people and it is here that the real networking comes in. Making yourself known and likeable to the musical community should become a priority. Join the local branch of your trade association. In the U.K. there are the Music Producer's Guild and the Association of Professional Recording Services, to name two we'll explore later in some detail in the book. Organizations such as these frequently hold lectures and events to which you can go along and rub shoulders with your kind. For U.S. based associations, see Appendix F-1, The Tape Store.

Networking is not simply about being in the right place at the right time. As with any industry, you need to gain the confidence to make an impression, which will start with a few introductions where you'll need to have your elevator pitch—something that quickly describes what you do and who you do it for. It's a crude business term, usually reserved for those people outside of your industry. Nevertheless, you may choose to prepare one for your industry. Keep it short and make sure you have a business card to hand if you want them to remember you. As we'll cover in the next section, it's important to get your brand right and to reflect the work you do.

If networking in person is not something that comes naturally, take some tips from Will Kintish (see sidebar) who knows a thing or two about the art! In his book *I Hate Networking* (2006, JAM Publications), Kintish explores the fears and countless solutions to getting through the barriers of networking in business.

WILL KINTISH ON NETWORKING

Don't network? Don't succeed!

The word network often strikes fear and dread into people or they immediately have strong negative thoughts about it. Manipulative, scheming, selling are just three words associated with this all-important activity. The irony is we all network from the time we start to talk. In my view, it is simply building relationships, either new ones or reinforcing existing ones. "I'm going networking once this week" implies you're going to spend the rest of the week sitting in a darkened room with no access to the outside world. It's just communicating, be it face-to face, on the phone, or more and more through the computer. Social networking or social media (the phrases can be interchanged) is becoming the dominant method of communication. As someone who acknowledges he is a little quaint and old-fashioned, I fear for the future of interpersonal communication. There can never be a substitute for attending gigs or industry events, no replacement for the smile, the eye contact, and the reading of the body language.

Building relationships

The three key steps to building new relationships are

1. Get to know more people by attending more gigs and events.
2. Start to get them to like you and build rapport and affinity
3. Continue past Step 2 and build trust to create long-term meaningful, sustainable relationships

I believe the reason the word attracts such negative views is because many people simply don't know how to do it effectively and, more importantly, ethically. This can result in rude and discourteous behavior which includes people being too pushy or, if they realize you're not the person useful to them, they begin to look around the room or over your shoulder.

Remember that the type of personality someone has does not prevent them from being genuine or polite. People do have a choice.

Why attend gigs and industry events?

You go to

- Raise your own profile
- Gain useful information
- Understand your marketplace
- Meet people who may one day collaborate with or work for you
- Meet key industry people and decision makers
- Get to know what others do
- Get others to know what you do and offer
- Help others with their musical challenges

This list is not exhaustive. Consider for a moment all the potential opportunities you will miss if you don't go. I say regularly, "If you don't go, you'll never know."

Fears and concerns

Apart from the negative press networking has created, we all have the basic primeval fears when it comes to walking into a room full of strangers.

1. Fear of failure
2. Fear of the unknown
3. Fear of rejection

Let me share with you some tips and ideas to help you overcome those fears which should then give you more confidence to attend more events.

Fear of failure. You won't fail when you spend time asking good questions, listening carefully, and following up in a professional manner when an opportunity arises. When you focus on the other person and show interest, people start to like you quickly. You need to be genuinely interested and when the conversation comes to an end, move on in a polite manner. When you hear something you don't understand, ask them to explain in more detail what they mean. Many people love talking about themselves and showing they know something you don't. This will endear you to them. You only fail when you don't turn up, you do too much talking, you are impolite or, in my view worst of all, don't follow up when you think you could move the relationship to its next stage. When you ask permission to contact someone after an event and they say 'yes,' no one can accuse you of pestering or annoying them.

Fear of the unknown. Walking into a room where you have never been before and knowing no one is scary. Even I hate walking into a room full of strangers so I always avoid it. How? Simply by planning my day carefully and arriving early. I have presented for a decade asking tens of thousands how they feel and I can confidently say 98+% of people have similar fears.

Every room you have ever been in and every event you attend in the future is always formatted in exactly the same way. There will never be more than six formats. There is the single person standing against the wall. Couples stand in open and closed formats as do trios. Then there are the scary groups of four or more. My advice is to avoid the closed-formatted groups unless you know someone in there. Approach singles or open groups with a smile, good eye contact with a phrase like "Please may I join you?" or "Please may I introduce myself?"

Fear of rejection. Most people won't make that first move for fear of rejection. Fear, for me, is an acronym; it stands for False Expectations Appearing Real. We walk into that room full of negatives. "No one will talk to me," "I am not going to be interesting," "What if I'm judged as uncool and found wanting?" Most people are friendly and polite so leave those words behind when you arrive.

All I say is believe in yourself, walk in and tell yourself you are a nice person and remember most other people are nervous. If it is an industry event they want to meet you, just like you want to meet them. The chances are they will be feeling the same kind of insecurities that you are, even if they are more experienced or further up the professional ladder.

Warning. You will meet the rare lesser-spotted R.I.P. This is the Rude Ignorant Pig who will reject you, walk off, or just ignore you. Give them short thrift when you encounter that behavior and move on.

Will Kintish is a leading UK authority on effective and confident networking both offline and online. His company runs workshops on both face-to-face networking and LinkedIn. He has become a leading expert in getting great value from LinkedIn and regularly runs web-based seminars on all its aspects. *Visit www.linkedintraining.co.uk and www.kintish.co.uk for further free and valuable information on all aspects of networking. Or call him on +44 (0) 161 773 3727.*

The process of networking

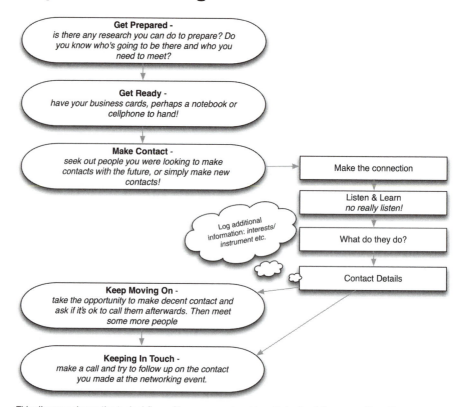

This diagram shows the typical flow of in-person networking. Make the follow-up call and keep in touch!

As you go to gigs in your spare time (still "on duty" naturally) and studios for work, you'll meet countless like-minded people along the way. These people may even become future collaborators or clients. Making a connection in the first place is down to confidence and the networking skills.

Many freshers (a.k.a. freshman) students we have had the pleasure to teach over the years have struggled to learn the art of networking, dismissing it as something unnecessary, both for a career and for their degree. We suspect that it was too uncomfortable for them to approach due to their shyness and perhaps lack of confidence. Overcoming these traits is never easy for freshmen students or for some seasoned professionals! We will not suggest this book is in any way the solution. However, there are tricks to the trade.

Making a connection at an event, or at the bar, will be something that happens, and seizing on the opportunity will be up to you. The opportunity might not present itself there and then in the conversation you have, but you should seek to exchange contact details if you feel they may be of interest to you, or indeed you may be of interest to them but they've not realized it yet.

Exchanging personal details in some ways has never been easier with the advent of the website and the cell phone but we still fumble around for that business card, something we'll discuss a little more later. Keep your business cards to hand at all times. Be prepared to add extra information to the back of your card for them to remember you by. This can prove really successful in business, we're told. So carry a pen (and paper too).

Being a producer, the way we're painting it here, might look very "business" and not "studio." Well, it's true, being a producer is as much about business and business standard skills as it is the glamour of the recording session. For that three-minute song of perfect sonics, many weeks may have been invested to produce it. There will be boring bits too. One of these bits to some might be the core industry standard business habits along the way.

So you've made your connection and are talking to someone you wanted to meet at a gig, or industry event. Now what? What do you talk about? You'd hope that the conversation matter would not be too problematic in the first place, given that this person you've wanted to speak to is perhaps a like mind, perhaps a budding engineer or assistant, or indeed an artist you'd like to work with in the future.

First, do some research if you're specifically going to meet someone. Just as Tommy D researches what his fellow songwriter/artist has been doing, it is worth getting on the Internet to check what your key prospective contacts have been doing. They'll be impressed if you can ask them about something they've done that not everyone has picked up on. It shows you're interested. It shows you're ready to listen!

In this first conversation you might want to get to talking about business straight away, especially if you know what you want from them—to record their next album, for example. However, many of the people you will meet will be simply by chance and not intended contacts. Talk to them about the music industry, and be sure to ask them lots of questions. It is no secret that people love to talk about their favorite subject: themselves. If you let them do this, you'll soon be in favor with them.

Once you have their attention, you can ask them some more prying questions about what they are doing in their careers and how that might help you, or how you might be able to help them. Dependent on their responses, you can make a mental note for the future of the type of music they like, the instruments they play, or the studio they like to work out of. In most personal information managers such as Outlook or the Apple suite of apps, you can then search for, for example, "bass" and find a list of bassists you might need in a forthcoming session. Being organized is key!

Back to the gig or bar; it is likely that they'll not give you a business card, as not everyone may be as organized as you, but use your iPhone or cell phone well. Most of these devices have the ability to store some notes in addition to the usual name and number. Place all relevant details in this for the future. If you're brave enough, you could even take their photograph so you have a visual record. Be warned, most people don't tend to like this!

Craig Golding

 Craig Golding
Course Leader of Music Production -
Music Production Department
Leeds College of Music

work 01234 567890

work c.golding@whatismusicproduction.com

homepage www.whatismusicproduction.com

Note: Bass, Guitar and Keys
Music Producer & Engineer
Live Sound Engineer (big venues!)
Works in education and looks for pro guest speakers some
times

Updated: 10/07/2010

Edit

Be smart. Use the notes feature in your cell phone to take down pertinent information and even a photo of the chap you met at a show last night. You might need a sound engineer of repute one of these days.

Taking a note of the people you've met will be of use to you as it's good practice where possible to follow up any connections, if not all, that you feel are worthwhile. If you've made the connection and said that you might follow up with a call or an email and they've not disagreed, get in touch with them, as Will Kintish suggests.

Making connections such as these increases the network of people you know and the people you can interface with. This will, in time, be a resource of great value as you build your career.

KEEPING IN TOUCH WITH YOUR NETWORK

Keeping in touch with the network you have developed over your career can be a tricky thing to maintain. You get busy. People get busy. And before you know it you're driving to the studio thinking, "Whatever happened to Johnny? Did he get that deal from that major?"

The age-old synchronous "picking up the phone for a chat" technique can still be the best as it keeps things personal and keeps fluid and personal communication open. However, it is a game of chance unless you've agreed to meet at a certain time. In the old days before cell phones, you'd phone Johnny and he'd not be there, but you leave a message with his roommate. Johnny calls you back but you're en route to the studio. You don't then pick up the call until the wee hours of the morning. These connections can get lost.

The cell phone is a major change in the way in which we now communicate due to the ability to send text messages (SMS) to each other. The other major change is the widespread adoption of email. Keeping in touch is not so synchronous as it once was, and decent and meaningful messages can be kept with people in your network.

A negative of all this instant-access technology is, of course, the junk. Getting your message through is one thing, but of course the noise then prevents you from spending the time and consideration you once might have done writing a letter in the old days. We're not suggesting this communication method is dead, but people are now getting very smart at informing their networks of what they are doing and how they are making this happen. Some fantastic, widely adopted tools have cropped up recently to ensure that, like your contacts in your network, you're keeping abreast of activity as well as informing everyone of your latest projects and thoughts.

Social media

One of the ways you might best connect with people you meet these days is by linking to their Facebook account while you're there, or look them up later, which is fine too. This realm of what is known as *social media* is becoming common for musicians to manage their networks.

Meanwhile other networks that are very suited to music, such as MySpace, appear to continue to be very popular for the sharing of unsigned (and signed) acts' music. Social media solutions such as these are proving to be very popular networking sites for the musical community.

The more professional of the crop is LinkedIn. Despite not being particularly musician or producer friendly, it is a corporate-focused tool and many a label professional can be found lurking there. This can be good if you wish to make some pertinent connections in the majors or wish to find an avenue for a new act.

Keeping your networks going on all these sites can be hard work, although these days there are ways of managing them all at the same time using one feed, as we'll discuss in a little while. Despite the fact it might be time-consuming, many a band has apparently been found for production through MySpace and many a job has come to a studio through its page on one of these social media platforms.

One of the newer, and more prevalent means of social networking is Twitter. Many producers use this to announce the work they're doing or things that they find interesting. It can be a good way of keeping up with your immediate networks or to use as an active news feed attached to your website.

Using these tools will ensure that you remain in the consciousness of the networks you maintain and hopefully when the time is right, one of your networks will pass you some work.

SOCIAL MEDIA BY WILL FRANCIS

Will Francis, a social media expert and former editor of MySpace U.K., speaks about how social media can really get the message across. "Social media is based upon four key pillars currently: YouTube, MySpace, Facebook and Twitter."

"The power of social media and getting noticed is not just keeping all these sites up to date, but making the connections: connections between each pillar, and also making personal, and real connections with people. Social media reflects real life and it's still real people behind it all. Talk to people directly, replying to their tweets, comments, videos, etc. Build conversations but don't spam or become annoying.

"The old maxim about 'who you know' being at least as important as 'what you know' is as true now as it ever has been. Only these days you have powerful tools designed specifically for building and communicating with large networks of contacts. It's not complicated or hard, but it does require a large investment of time and a good dose of common sense about what other people like and what will irritate them. The 80/20 rule about balancing personal, conversational content with self-promotion respectively is always a good guide."

Speaking of his time at MySpace, he recalls a time when the site was relatively new "…and it could propel a band seemingly into stardom. The truth is that bands like the Arctic Monkeys already had a fantastic following in Sheffield, UK and MySpace provided a catalyst for some further buzz.

"MySpace is no longer the exception, but the rule for artists. It would be odd for an artist not to have a significant presence on most of the four pillars, especially MySpace. Where an artist's online presence comes into its own is where it can maintain interest for a band between albums and on the long tours when the label isn't running any promo for them. They can connect with their fan base, providing snippets of personal content like blogs, photos, and videos, keeping fans engaged."

Client record management

Client record management systems or contact record management systems (CRM) are elaborate databases which act as repositories for a whole host of data about the work you do and the clients you hold. Most businesses have one as it accurately describes how you work with your clients, who has spoken to them, and will even bill them for your time.

As we'll explore a little more in Chapter B-3, Being a Business, you can see how such systems relate to the studios in which you may work, but CRMs can also be extremely important for any professional anywhere. To have a sophisticated handle on the work you're doing and the clients you have at any point is an extremely valuable thing to do.

Many producers again might think this is a step too far choosing to use a PIM (personal information management system), but it is likely that every producer or manager who has more than one client on his or her books is using such a system. Also, if you become a producer who has a team of people managing your time and sessions, you might choose to employ the services of one of the CRM systems available.

PERSONAL INFORMATION MANAGERS (PIM)

Personal information managers or PIMs is the term we used to call our address books, calendars, and perhaps emails, all in one. A widespread example would be Microsoft Outlook or Entourage for Mac. In either case, these are fantastically richly featured PIMs which can, for most people, manage their lives, especially with the link-up now with mobile devices.

PIMs have some shortcomings which are becoming less and less of a hindrance. One is the sharing of information with colleagues. This is fine if you have invested in an exchange server (the corporation's server behind Outlook) or you've spent the money to rent an exchange space from a hosting company. But if you're on your own and working with perhaps one other person (a part time PA or admin support), an exchange server is overkill.

For this small kind of link-up there are fantastic solutions which actually are free, such as Google's collaborative tools (in fact, this book was written collaboratively using Google Docs!). As a producer, you could employ a number of Gmail accounts and through this calendars with some documents to make a powerful PIM.

However, despite the fantastic, and free, options Google provides, it is not a professional CRM working toward the goals of the average music producer. Before we delve too much into the alien world of CRMs, we ought to dwell a little on the average PIM suite.

For Mac users, which many producers are, the suite is iCal, Mail, and Address Book. These three apps are all interlinked and work fairly well as a PIM. If you were simply organizing your life and a few sessions, then this is absolutely fine, but you might wish to also consider the employment of Entourage (at the time of writing Microsoft plan to bring Outlook to the Mac). It might be a wise adoption, especially with its Project Manager feature. This allows you to set up different projects per artist and session which might make tracking a little easier.

One thing these PIMs will not do is organize your billings and also any other aspects to do with your clients, which CRMs typically can manage for breakfast. Let's elevate to the smaller CRM and have a look at what it has to offer.

One-man-band billing and CRM

Should you decide as a producer that a PIM is good, but you're losing track of your word processed invoices and your spreadsheets, perhaps it is time to look into the smaller CRM solution.

This is not to go the high-tech end, which we'll discuss shortly, but if you just want some sensible software to manage your own version of CRM and some billings. A really neat solution we've come across is Daylite (available for download at *www.marketcircle.com*). Daylite integrates well with Apple Mail with the additional Daylite Mail Integrator. It is also scaleable up to multiple users as your company grows.

Daylite is a solid CRM that can run locally and improve the way in which you interface with your clients especially with Mail Integration. Another aspect to CRM work is how it manages your billing and invoicing. Marketcircle's Billings software is also very good and integrates to Daylite, making a relatively low-cost, yet effective, Mac-based system.

Daylite from Marketcircle is a good general CRM system for the individual or small firm. With some iPhone integration and with some additional bolt-ons, it becomes fantastic for Mail integration and for billing.

BIG BOY CRMS

CRMs are by their nature scaleable systems. Some large corporations will wish to keep a file per customer and all their correspondence. You'll recall the telephone conversation with customer services where they say "please hold while I bring up your records." They're entering in your customer number, or last name, in a hope that there are plentiful details of the complaint you made last week. This is the CRM at its best, working to serve the whole customer service team, and to ensure that the company gives a parity of response to the same customer from many different angles.

That's at the top end, although you might, if you escalate to having a number of people working for you, wish to have the same shared knowledge. Even as a one-man band, you may still wish to employ the services of a small CRM system locally on your Mac, especially if it offers you some semblance of order among the myriad projects you're engaged in.

There are those CRMs that are truly meant for the professional markets, such as ACT by Sage and Dynamics CRM by Microsoft. These both are server-based intensive systems for larger corporations. There are web-based systems such as SalesForce, which is highly thought of.

MUSIC SPECIFIC CRMS

None of these systems really pander to the music production community. For CRM work, you could design and operate your own system. Many geekier producers and engineers we know have used Filemaker to design and develop their own CRM style database. The geeks among us will know what this is like! However, there are those who have taken this to a whole new level.

Using Filemaker as a platform on which to write a database, companies like Apogee during the 1990s, with the help and idea of the famous engineer, Bob Clearmountain, created a system called Session Tools. This was one of the first CRMs specifically for the studio. It would allow you to keep all your contacts in one place, alongside both billing and also total recall notes (notes of all the settings on your favorite kit for each session for recalling at a later date). This was in those days a fantastic introduction to what would be developed later by other companies such as AlterMedia and Farmers WIFE.

Next up is AlterMedia's Studio Suite, which has championed the studio database on the Filemaker platform. As you can see below, this is a comprehensive suite for the studio and perhaps producer also.

Studio Suite, unlike a common CRM, has the ability to not just simplify your office but also manage your studio and organize your tech. This is a powerful database that manages billing, bookings, and patchbay labeling.

There are other systems you could come into contact with when booking, or indeed choose for your family. Studios such as Strongroom and AIR in London choose a system called Farmers WIFE which can also integrate with staff members' iPhones, meaning that the studio schedule can be with you at all times, if they'll let you access it!

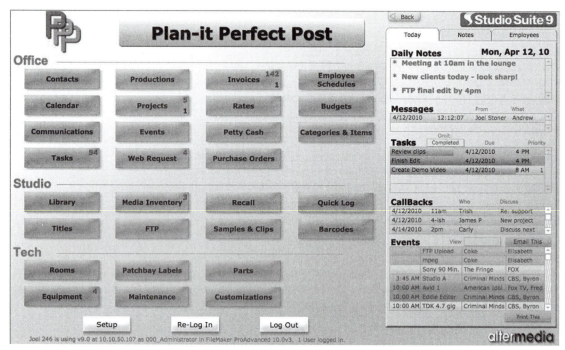

Studio Suite is one of the major studio management tools around and should you get your own facility, this could be a sensible CRM. For many of the studios you may work within this could also be their choice.

CONCLUSION

We hope that this chapter gives some insight into the being of a producer. It takes social skills, musical creativity, and personal integrity. Add to that mix some organizational skills, which we'll look into in a little bit more detail in Chapter B-3, Being a Business and later in Chapter C-3, Project Management.

CHAPTER B-2
Your People

"It's not what you know, but who you know."

Anon

INTRODUCTION

The music industry is not only built on a strong foundation—music—but also its greatest commodity—people. As a producer, it is the people you'll meet, collaborate with, write with, engineer for, and do business with that make up what Napoleon Hill called a person's "mastermind alliance." Hill, one of the first gurus of personal achievement, in 1937 asserted the benefits of a group of like-minded people "working in perfect harmony toward a common definitive objective." Many among you reading this will think we've gone mad, but there's some synergy between Hill's thoughts back in the 1930s and the music producers of the 2010s.

The team you use, amassed from years of collaboration, fun and business, will become the group that mastermind you to success. No producer chooses to work with a less able drummer or engineer than he or she has to, unless there is some loyalty that keeps connecting them.

Of many industries it is said "It's who you know, not what you know" and never more aptly could it be applied to an industry than to ours. Making the best of your team to deliver on a project is part of the day-to-day work of the producer. It is important to select the right kind of team to work with that share the balance of solid working ethics with a sense of open innovation, making you open to possible experimentation.

There's an excitement we all feel when we're in the studio capturing a "moment" or writing music that moves us. This excitement is often where the options to choose different outcomes present themselves. Inspiration comes at certain unplanned times and we require the flexibility to embrace these serendipitous opportunities as they emerge.

This chapter focuses on the people the average producer will come into contact with and how they fit into the music production jigsaw. Starting with the artist,

What is Music Production. DOI: 10.1016/B978-0-240-81126-0.00004-4

moving through to any personal assistants you may hire, we discuss how the mastermind alliance for the producer connects and thrives.

Sir George Martin at a recent charity event we attended responded to a question from the audience about the importance of the record producer and suggested that the order might look something like the following, finishing his response with "to hell with the producer" to an audience of laughter.

1. *Composer*
2. *Performer*
3. *Producer*
4. *Engineer*
5. *The rest!*

Martin identified that without the music creation, there's not much for the performer (artist) to go at and that the producer is much farther down the list of important people. We'll begin by adopting this order.

Composer(s)

The composers are of primary importance to the process. Their music generation is the germ of the success through which the interpretation of the band and the production team provide vital assistance. Without the song, whether internally (by the artist) or externally written (by another writer), the production is not going to get very far.

Working with songwriters, whether internal or external, can prove difficult. Often conflict within the band or the production team can occur as differing opinions come to the fore. As the producer you will try to smooth over disagreements where possible and bring the music to the best fruition possible.

This can be difficult in situations where the songwriter is resistant to external influence, stifling the production of the song. Imagine a situation such as this where you, as the producer, can see clearly that improvements can be made to a song to make it a hit, while the songwriter is remaining resistant.

Other conflicts can occur between the songwriter and the band along the same lines inasmuch as the band wish to develop, or get in on the writing process, and are given the cold shoulder by the writer. As producer, it is perhaps likely that your inclination will be to accept and try out, wherever possible, every idea that comes your way.

Many producers do not see themselves as the be all and end all in such situations. Of course, many producers have to be decisive and lead the way, but most we have spoken to agree that they are rarely dogmatic enough to suggest their ideas are the only way forward. In many of our conversations, the words *facilitator* and *decision maker* are commonplace.

However decisive they are, they will have to deal with conflict inside and outside the studio as we'll discuss in Chapter D-1, The Session. Needless to say, the producer's management of these situations is paramount.

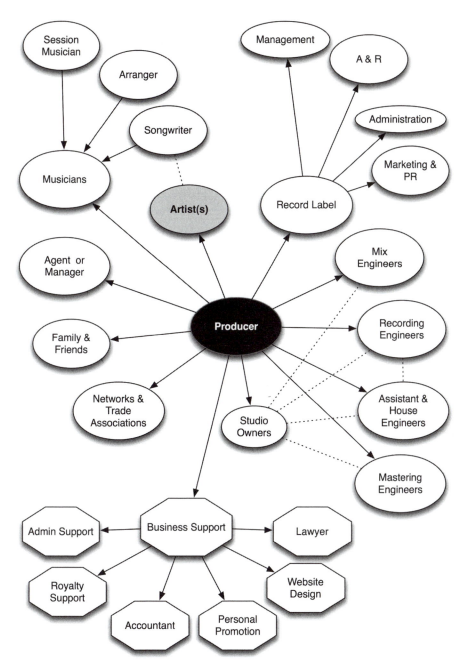

Being a producer is a lot like being part of a larger company or corporation as this diagram shows. These people make up your *mastermind alliance*.

Nevertheless, a shared view from all the producers we have spoken to in preparation for this book is that the song is the most important thing in production. As such, the composition and the composer should be developed and understood to assist the production.

The artist(s)

It's needless to point out, but artists are some of the most amazing and talented people you'll work with during your career. However, as folklore would have it, much of this tremendous talent can be wrapped up in either variations of anxiety or arrogance, much of which you have to navigate through to complete a project.

Developing relationships with your artists will vary. Producers can and do develop decent, sometimes lasting, friendships with the artists they have the pleasure to work with, but many do not. The studio can be a frenetic and often boring place to operate and, at times, tempers can run high. Many such relationships remain professional when in the studio, where the artist is ultimately the client, but outside of the musical sphere, the friendship remains. Other examples can be where artist and producer simply work together because they have to, and go their separate ways at the end of the sessions. Plus, of course, there are those sessions of folklore which describe Armageddon within the control room.

The key as the producer is to manage those sessions and ensure they work successfully. Becoming a master of managing issues, egos, or sensitivities in artists should, in time, become second nature. This is not to suggest that all artists have issues about their work; they do not. Many artists are confident, affable, humble, and open to development, making them a pleasure to collaborate with. But what for those who do not neatly fall into this category?

Let's be honest here. It's difficult. You win some and perhaps lose others. Working with people in confined spaces such as the studio, and for something that is obviously precious to them, is naturally a place for differences of opinion. Learning to manage these differences is one of the most important roles we believe a producer may be required to master.

It is all about trust, as we mentioned in Section A. Understanding the artist and how much the music means to them is also about empathy. As many producers are either musicians who have written or recorded material or engineers who have seen it all before in the studio, you'll be well-versed in what the artists are going through. We cover the studio session and interactions in much more detail in Chapter D-1, The Session.

On the flip side, there are those producers and artists who see the importance of developing long-lasting professional relationships. There have been many different examples of this. One such example is that of Van Morrison with producer/engineer Mick Glossop. Glossop has worked with Morrison for some 25 years and is clearly a trusted pair of hands on any new material. Other examples can be that of Toni Visconti with many of David Bowie's albums and Hugh Padgham's long stint with Sting, among many more.

Gaining trust such as this over many years is something that can be mutually beneficial both in terms of business and creativity. Many producers report the almost telepathic communication that can occur between artist, engineer, and producer when the sessions are going well and they "click." This can be effective, efficient and fantastic ground for new creativity despite the seeming familiarity between all involved.

Seal writes in his sleeve notes to his 1994 Grammy Award-winning self-titled album of his desire initially to seek out a new producer:

> "During this time [the album has] been through quite a few changes as I'm sure you can imagine. Different songs, different locations, and even different producers, which brings me to what was perhaps the most significant turning point in the recording of this album. Being the type of person who doesn't like repeating themselves, I thought it would be a good idea to offer a new sound and hence a new producer, but what I didn't take into consideration was the rapport that Trevor and I had built up during and after the first album, and that he and I were both thinking along the same lines. He is in my opinion the only producer for me."

Seal stuck with Trevor Horn for another two albums until working with Stuart Price on *System*, but appeared to move away from what he'd been used to and once upon a time praised. Perhaps Horn's not the "only producer" after all?

It is clear from this change of heart that an artist's creativity can lead them down a different path. Much of what they write is fuelled by a number of influences, not only internal ones from the song, which may lead them to a new producer, but also from external influences such as the modern times they find themselves within. Things change. People change, and we are all, especially in the music industry, painfully aware of this. Some producers are indeed flavor of the month and as such artists will follow in search of success. Fashion dictates much of what the populist end of popular music does and how it should react. Producers regrettably can become popular overnight and then are looking for a new career a year or two later.

So what about the producers? What we hear time and time again from the people we have had the pleasure to speak to in preparation for this book is that it's the "music," the "song," or the "artist" that is the most important thing in making a record and, in this instance, the artist must be given that priority. These will be your closest co-workers in the studio alongside the engineer. Getting the relationship right will ensure a pleasant experience within the studio, good music, and a successful album. What constitutes success is naturally up to some debate as to whether this is from sales (and thus income) or whether the objective, or vision, has been realized for the artist. In either of the situations described above, the producer is intrinsically important to the process, working with those most valuable of assets, the "songs" and the "artist."

The Producer

Naturally, you're important; you are the producer. Actually, Haydn Bendall, former Abbey Road chief engineer, says "we're not that important and in comparison to the artist we're very replaceable!" preferring to interpret the abilities of the artist and the quality of the song.

Anyway, in this part of the chapter we'd be foolish to discuss at length the producer, as that is what this book is mostly about. However, we wish to focus on the relationships and how these are made and managed.

As we established in Chapter A-2, Analyzing It, many producers would say they are opinionated and are paid to be so. But they are not without a great deal of humility. Some producers we have had the privilege of interviewing are some of the nicest and most personable people you could hope to meet and work with. However, many share traits insomuch as they want to get on with people and collaborate. They can often be natural networkers, asking questions and making connections for the future.

We spoke about the art of networking in Chapter B-1, Being A Producer. Suffice it here to say that developing the right network, or mastermind alliance as Napoleon Hill would have suggested, will ensure a productive and valuable set of collaborators to work with. Now many producers are generally not too shy and can go out there and do business in the networking world. Some are, of course, more shy and overcome these issues in other ways, perhaps through raw talent, or simply the ability to listen and interpret an artist's ideas.

FELLOW PRODUCERS AND CO-PRODUCERS

It is rare that you will work closely with other producers in the studio or on the same act, but it does happen. It happens when, for example, you're one producer on an album containing songs produced by someone else. It may also happen where one producer has begun a project with an artist and for one reason or another, the work has been shifted to your in-tray.

It would be quite rare for you to work, or collaborate with, another producer on the same track, unless you chose to. However, it would be perfectly natural to have a co-producer, such as an artist or band member who works closely with you. Finding ways to collaborate, share, experiment, and agree on courses of action will be the key to a successful record.

The studio owner/manager

You will develop relationships with studio owners or managers and their staff as your career develops. You'll want to book one studio for one recording job, another for the orchestral tracking, and perhaps another for the overdubs and mix. These different venues will have each of their own management structures based around the studio owner or manager.

Your relationship might be with those studio managers who are more involved in the day to day activity of the studio and the music industry. Others, such as businessmen who may be less active on the music side, or indeed investors who may be silent owners, you'll have less to do with.

As with any relationship, it needs developing. The studio owner will see you as a client and will treat you with respect and will, certainly in the current economic climate, want your booking. You'll get to know these people if they're involved in the day to day running of the studio, but the larger the outfit, the more likely you'll be discussing your requirements with some of the studio staff.

For those studios whose owner is either an investor, or otherwise engaged in other business most of the time, a studio manager, or director, will hold the reins to the studio complex. This person will assume full management for the personnel and smooth running and development of the studio. Again, a good relationship with these positions of power within the studio will be beneficial for you should the option be available to you.

The more you book at the studio and the more you're part of the furniture, the more you'll get to know everyone and be privy to special rates that might come your way if they can. However, the starting point for your bookings into the studio and developing a longer term relationship with the studio will come from the studio staff with whom you'll have most contact.

Studio staff

The studio complex, certainly larger ones, will have sometimes up to 20 or more staff to keep things ticking along nicely. Obviously there are the studio owners, directors, or managers which keep things running smoothly. There are also a whole host of other staff which you'll come to relate to as you move from studio to studio, ranging from the booking staff, house engineering team, maintenance engineers, to the all-important in-studio catering!

BOOKINGS

Each studio will have one main person (or more than one, dependent on size) through which studio time is booked. You'll get to know this person very well and it would be well worth your while looking after them. They will become your conduit for booking time and will be key to you planning your sessions well into the future during pre-production.

Most studios will employ a database driven system to take details and make your studio booking. This is part of a wider database type known as a *client record management* system (CRM), which we introduced in the last chapter. In the world of commerce there are many systems used in all businesses for managing your data, which employee spoke to which client and what was said. It will manage all the email promotions and things you've purchased from that marketing campaign. It's all pretty clever stuff and for the likes of large corporations, it's vitally essential.

CRM systems in the recording studio are slightly different. Of course, one of the above could be tailored to become a music facility specific tool, but there are already systems in place as we discussed in the last chapter.

HOUSE ENGINEER(S)

In the early days of the studio, the house engineers were the only people available to you and it would be extremely rare for a freelance engineer to come into a studio, removing the house engineer from his seat.

Part of this culture was born out of a need to understand the equipment of that particular studio. Most early recording equipment was developed by the studios in which they were housed and used. For example, EMI at Abbey Road developed much of its own equipment, or indeed commissioned people to develop things such as the RS124 compressor. Trident, another famous U.K. studio from the 1970s, was also renowned for developing its own equipment, and in this case their mixing consoles.

These are some later examples of this trend, as very early studio complexes had to build all their equipment pretty much from scratch or experiment with things to make new and exciting recordings. It would therefore be akin to intellectual property rights management to keep hold of your staff for as long as possible, because your inventions and ways of working were not common knowledge and standardized as they now are. We all use Pro Tools, and similar solutions now, but in the early days proprietary recording devices and formats would adorn different studios.

The house engineering teams in these early days were large in size, given the mechanical nature of the recording process with large recorders, tapes, and reverb chambers. It is only now that engineering teams can be reduced to one producer/engineer given the advances of technology.

In line with this has been the shift of activity for the house engineering team at any studio complex. Today's house engineering team will often consist of a head engineer who could step in or be hired if the artist was not bringing their own engineer. Alongside the head engineer would be perhaps a number of assistants that incoming engineers and producers could call on to help them during sessions.

ASSISTANTS

(See also Freelance Assistants, below.)

Already introduced earlier in the book, assistant engineers are a breed of engineer whose role has changed somewhat over the years. Again returning to the pioneering early days of the studio, the head engineer, or main engineer on the session would manage the console, balances, and so on, but he would have at his disposal a large team of assistant engineers who would manage the recording process, often mic positions, and generally ensure the technical side of things was manageable.

As part of this team, there would be a role called the tape operator. The tape operator, often shortened to tape op, would manage all aspects of the tape machine, ensuring the open reel analog tape would be clean and set up to its optimum performance. Not only this, most tape ops would be required to manage the transport controls (stop, play, record, etc.).

Despite being an incredibly responsible job, especially in the days before the undo feature, many tape ops were seen as the bottom of the pile and may have been asked to run all sorts of errands for the band and engineers.

These days, you might employ a Pro Tools operator, but more often than not the main engineer or assistant would manage this aspect of the session. So from a few assistant engineers many years ago, the roles have eventually reduced, partly due to the advance of technology. Assistants now perform a whole host of responsible functions in the session and can also be freelance, as we'll later explore.

Assistant engineers, given their knowledge of the region and the studio can also be valuable in ensuring that mics, equipment, and supporting items are located in the region for the engineer or artist in a timely fashion.

In the traditional model, most assistant engineers will have started in another supporting role such as the receptionist or even have started in the studio complex as a volunteer or under work experience.

STUDIO SUPPORT STAFF

Every studio of a larger size will have a number of support staff to ensure the smooth running of activity. These staff can range from the bookings person, as we discussed earlier, through a whole host of possible roles that may differ in each studio complex. We've put together a rough list for all the roles you might find for those larger studio complexes.

> Studio owners
> Directors
> Studio manager
> Studio bookings
> House engineer
> Assistant engineers
> Maintenance engineers
> Administrative support
> Financial and accounts staff
> Receptionists (can be future assistant engineers)
> Catering staff
> Cleaners
> Other staff (including volunteers and work-experience personnel)

Naturally you will come into contact with these staff in your time in the studio. You will have more liaison with some than others, such as the receptionists and catering staff, but these may not have an overbearing outcome on the success of your session (unless you upset them, of course).

More obviously, you'll have some connection with the administrative and accounts staff as you'll be billed for your time in the studio and for the equipment you've hired from the stores. As with any business you'll find your key contacts with which you speak about aspects of your work. Keep these contacts up to date, especially if you're using the studio a lot. They'll keep you posted about the politics and inner workings of what is going on in the studio and the business that is passing through it. Some of that information might be very valuable moving forward.

ENGINEERS

The team you surround yourself with, should you need one, in the studio will be of paramount importance. This will be especially so if you're not going to get your hands dirty on the faders (or mouse and keyboard these days). As we've previously mentioned, many producers are engineers in their own right and as such would not contemplate passing the reins of the mother ship (a.k.a. the mixing desk) over to anyone else.

However, should you be a musician-producer, such as Trevor Horn, highlighted earlier, you will surround yourself with a number of talented engineers to enable your workflow. There are some excellent benefits to this, as you can work on more than one project concurrently, as I dare say someone as prolific as Horn does from time to time. More importantly is the benefit that you can leave the session for as long or as little as you want. While this might appear detached, it also allows clarity and objectivity at the many stages, as the engineer can be left to simply get on with tracking or mixing. The producer can then attend with fresh ears and provide clear advice regarding the mix. Robert Orton, who spent many years working as Horn's engineer, says of the times he got to produce, "It can be difficult to separate yourself from your engineering duties to think musically. The engineer in me can be distracting from the producer in me."

Choosing the right engineer will be down to performance and personality. The skill of the engineer is to capture and perhaps mix later the recordings. Naturally they'll be assessed as to their ability to capture great recordings. Also, personality is of course a huge part of it. Joe D'Ambrosio, producer manager for Tony Visconti among others, suggests that "producers have their guys [chosen engineers]… it's called having someone you're comfortable with just like a mate in life."

The ability to be cool and calm in the face of sometimes stressful activity can be paramount also. Being someone who the artist and you can get on with in the studio is to some more important than actually being the best engineer on the block, although it helps! Even if you are the best on the block, if you're difficult to work with, you will be less likely to get the gig. Phil Harding, engineer for Stock Aitken and Waterman during the 1980s, comments that the skills required are "character, attitude, and the ability to get into the right headspace," something that he describes as the "zone."

It can be common to employ a professional mix engineer these days. These specialist engineers focus on mixing and as a result become very honed in the tools of the trade, understanding how the music will translate to the record buying public. Tuning vocals, tightening up live drum kits, adding extra elements where required and permitted all point toward a glossy and highly professional mix. This practice appears part of the process these days alongside a producer when the sums are calculated for the project. To gain the services of a mix engineer such as Tom Lord-Alge or Robert Orton can send a signal to the label that the producer is right behind the project or indeed the label has the investment clout to get the best names in the business to preside over the track.

The same is extended to the mastering engineer who will further perfect the mix. There are a handful of mastering engineers across the globe who engage in much of the available work. Some producers will have a chosen few mastering engineers. Tony Platt has worked frequently with AIR Mastering's Ray Staff.

Freelance engineers

Freelance engineers have become commonplace in recent years working in many different studios for many different producers. The world of the music production professional is now mostly freelance, as the days of staff engineers at large studios are mostly over. Producers will call a freelancer they have worked with before to engineer their session before asking the house engineer in most cases.

For this reason, the freelance industry is where you'll find yourself, unless you choose to become a company in perhaps a collective of producers or as a production team. This mode of operation is something we discuss in more detail in Section D.

Most engineers will be the same and work individually, some managing themselves, while others will use managers or agents to manage their time, as we'll come on to later.

RECORDING ENGINEERS

The majority of engineers these days are freelance and will move from studio to studio, but there are a whole host of others that either own their personal studio or are associated with one. Engineers, like all creatures, like to work in their favorite place, and certain studios can have that lasting appeal. Some engineers such as Jerry Boys will always be synonymous with Livingston Studios as the renowned engineer he is.

You will work with a variety of engineers in your time as a producer, unless of course you prefer to take the controls yourself. In either case you will meet and make use of many assistant engineers and other music production professionals throughout your years in the studio.

Ensuring rapport with colleagues is important and you'll soon find out who you like to work with and who you do not wish to work with again. This rapport will be the thing that can make a session smooth and perhaps make you wish to collaborate in the future. In any case, as with any industry, you will soon develop a list of contacts with whom you like to work on projects, developing that sixth sense together as you record.

FREELANCE ASSISTANTS

There is a healthy set of assistant engineers out in the industry which support many professional sessions. These are the often unsung heroes of the industry, as they are those people who serve the session and do so in a truly selfless and professional manner.

While some may be freelance, many assistant engineers will be attached to a studio and provide an interface between the freelance engineer and the equipment and personnel of the studio in which they're working.

MIX ENGINEERS

The concept of the separate mix engineer is a relatively new one, with a whole breed of engineers making names for themselves solely in this area. Today we're all familiar with Chris Lord-Alge and of course Bob Clearmountain. Both now work exclusively in mixing. There are newer people to the mantle such as Robert Orton, who we speak to in Chapter D-2, The Mix.

It has become recognized that those people who spend the majority of their time mixing can provide an additional layer of objectivity and a sheen on a mix that other engineers may not manage. An example of this is an album where there are some key tracks (the singles in the eyes of the label) which have been sent to be mixed by a professional (and usually pricey) mix engineer, while the rest of the album remains in the hands of the recording engineer for the project. The difference can, at times, be startling and only then can you hear the difference and reason why mix engineers have become popular. We're sure you can think of a few albums like that if you've been listening for it.

This is not to decry the work of the recording engineer, whose mixes will be of a very high standard. Conversely, it is very likely that should the mix engineer go into the studio to track the band, they'd not make the same quality job of it!

You will need to decide whether to use mix engineers in preference to your own engineer or your own efforts. Alternatively you might submit your preferred mix and the label might still wish for one of the named mix engineers to remix the track for a more specified sound.

MASTERING ENGINEERS

Mastering engineers are, as we'll learn later, an integral and vital part of the process. Their work is the final, precious stage of the production, placing the "gloss" over all the tracks, to make them punch and flow on an album. The details of this we'll get on to in Chapter D-3, The Mastering Session.

From a producer's perspective, choosing a mastering engineer can be decided on many different factors. Is the mastering engineer known to you personally? Have you worked with them before? Are they renowned these days for the type of music your current artist is making? Or are the label, artist, or manager insisting you use one person over another?

Many engineers have their preference as to who they like to collaborate with, and such relationships can span a whole career. Many producers prefer to select an appropriate mastering engineer based on the kind of music they're producing and the market need at the current time. These may be from a select few top-flight mastering engineers across the globe, or it may be that there is one person you work with really well and believe he's brilliant and yet to be catapulted to the top team.

Certain mastering engineers will build up a reputation for particular styles of music for which they become renowned. They might not wish to be pigeon-holed in such a way, but for a period of time this can be good for business. As we'll discuss later in the book, this can be a similar issue for producers, whereby a producer will ride a wave of a certain musical style for a period and then lose prominence as the tides of popular music change direction.

Finding the right mastering engineer may be something you develop through trial or error over time, but you may choose to use different engineers per project based on their previous work. Developing this knowledge will take time listening to material already mastered and speaking to your engineers. Ultimately it will be trial and error until you find the people who make things sing for you.

Agent and producer manager

Many producers choose to engage an agent or manager to assist them in finding work and then managing it. Keeping up with the creative aspect of producing is not always compatible with the business support side of things, let alone the going out and finding work part. Being creative is the paramount thing and having to worry about minutiae can be distracting.

A producer manager can be vital in touting you to labels and artists in an attempt to get you your next gig. Conversely, artists and labels will no longer necessarily come to you personally, but will go through your agent or manager. This can become a fantastic filtration mechanism should you be highly successful and not wish to take risks with artists with a limited shelf life (unless of course their music has something that you can see will be a hit).

Asking a manager to take over business aspects of your work can be very liberating as the deals, the money, and so on can be taken out of your hands, leaving you to get on with production at hand.

This is not to suggest that the majority of producers have their own managers, but some may do and, then again, many do not. It is a matter of preference. There are those producers who enjoy the business side and feel it necessary to keep on the pulse of the industry, as it changes form and direction. This can make those producers the innovators of the stock, potentially finding new acts and placing them into prominence using new deal types and marketing strategies. On the other hand, there are those producers who are simply too busy or in too much demand to not employ a manager.

In this book, we will cover a little about the kinds of administration and business you'll possibly get up to as you rise up the producer ranks. Knowledge of all this activity can be essential to make informed choices as you go forward. You never know, you might be good at it!

JOE D'AMBROSIO

Production management came naturally to Joe D'Ambrosio. "I was always a pretty organized person growing up... friends... say 'it all makes sense. You were always the organizer...this is what you did'... and based on that I had the innate ability to just be able to pull things off."

Joe had many careers over the years and it was Phil Ramone who helped him become a producer manager. Joe had worked with Ramone first as his personal assistant and then later his head of production for his label. Meeting with Ramone later in life, Joe suggested that they could work together again, this time as his manager.

Joe's business, JD Management, has been in business nearly 10 years now and has grown out of this initial work with Ramone. Joe now manages 19 professionals including Tony Visconti and Hugh Padgham.

"The work is 70:30. 70% of it I have to go out and find, while the other 30% comes in to me for my client," says D'Ambrosio. "The world has changed and I need to go get work. The ones with the longer reputations and the more success, such as Jay Newland [co-producer for Norah Jones], have not stopped working for five years.

"I represent people," says D'Ambrosio. "I'm a manager to some people, but some would call me an agent in certain circles."

D'Ambrosio explains the work he does: "Not only do I oversee their business, look for work for them, make deals, close deals. I also buy homes, rent apartments, help Tony Visconti get a book deal and Hugh Padgham a speaking engagement." D'Ambrosio sees his role in far more of a holistic way for his clients rather than simply "closing deals."

Musicians

In this game of music production, it is the musicians that make its world go around. Without musicians, where would we be? There was a time just after records arrived on the scene that people said that live music was dead. In some ways it was, insofar as the piano stopped being the focal point for entertainment in the family home or the pub or bar. Recorded music took its place. Professionally, however, musicians were still employed, perhaps even more so.

The same claims came about with the introduction of more comprehensive drum machines in the 1980s, that drummers would be obsolete in a few years' time. This also has not happened. A drummer is still a key part of a band and most performances.

Therefore, musicians are still the key to making these things happen and, as a producer, getting to know the best in the business might be a worthwhile expansion of your address book. Producers at the top of their game working with single artists without their own band have the ability to choose the best performers available to them. Producers will have a list of special musicians, called session musicians, who they'll call on, or travel to in extreme instances, just to get the right take.

Musicians are not just the people placing the music on the tape, but are also the professionals behind the scenes such as the arrangers and conductors for that rich orchestral session on your artist's ballad. They're all part of the picture. Getting to know your mastermind team will be of great benefit!

SESSION MUSICIANS

A label comes to you with a singer songwriter with an album of material to produce. What do you do? Do you program everything in Logic, or do you play the material yourself with the artist? Or do you gather a host of highly talented and adaptable musicians to play the parts for you?

Should you choose the third approach, you're most likely to be calling in the services of a session musician. This strange breed are incredibly talented at their own instruments and are adaptable to many different musical styles and ways of performing. Session players will be able to read music, learn fast, provide accuracy with style and when required assist you in coming up with ideas for their parts.

Obviously session players are not just for the session, despite what the name suggests. They are also the team of people that the artist may call on to hit the road with them when they tour.

As with all walks of life, we like to do business with people we've used before, because they're good, and we trust and we get on with them. The same works for the session musician. Take Peter Gabriel, for example. He has used mostly the same session players for many years in Tony Levin (bass) and David Rhodes (guitar) alongside a small handful of drummers.

There will be times when you want something unique or to provide that extra dimension to your music. Then you may wish to call in a specific player or a specific instrument to add to this. One of the most frequently added session players to pop music outside of the traditional band format would be either the quartet or the orchestra.

SECTIONS, QUARTETS, ORCHESTRAS AND CONDUCTORS

Think of so many of the early popular music recordings and you'll often consider the band behind in the rock and roll era. Alongside this were of course the famous singers, which would enjoy the backing of an orchestra, or at least a smaller chamber sized orchestra. These players were to some extent the soundtrack to many decades of so-called popular music before the advent of rock and roll.

These days, in most popular music, the music generation is what we'll call here band-orientated or electronic. Band-orientated instruments usually are the drums, bass, guitar, keyboards, and vocals. Of course, electronics have taken their fair prominence in the past 40 years or so, especially with the explosion of dance music.

However, there are those sections of the recording in a session in which we'll choose to either use a Roland synth or a set of samples stacked up in EXS24 to provide the parts an orchestra would. However, you'll at times, especially if you have the budget, choose to go the whole length and go for a real set of players. Before you do this, you need to decide who is going to arrange the parts. Will it be you? Or will you hire an experienced arranger?

ARRANGERS

Before recording either a quartet or an orchestra, it is likely that you'll need to hire the services of an arranger. An arranger can have multiple roles, as you'll learn in Chapter C-4, The Desired Outcome (see sidebar written by Brian Morrell). These music production professionals can provide arrangements for many different instrument groups, such as brass bands through to full orchestras.

Recording an orchestra to add to a pop track can be immensely powerful.
Photograph courtesy of Mark Cousins, www.cousins-saunders.co.uk

Classical musicians are quite different usually to those involved in popular music forms. They have been highly trained to perform to a score. Their sight-reading skills are highly tuned and they do not often improvise. As a result, a score often needs to be produced and will need to be arranged. In the classical world, the writing of a score would be considered composition, but in the case here of taking a popular music piece and providing a score to accompany it, it would be considered arranging.

Arrangers are brought in to provide services of arranging the parts required should you want to place backing on your artist's ballad or a horn section on the retro disco track you're working on. An arranger is a fantastic ally in the world of music production and can really enrich a production.

Record company

The record company, or label, was traditionally the organization that would seek out the talented artist and through its staff bring together songwriters, musicians, and a technical team to record and then manufacture records. The record companies were, and to some extent still are, extremely powerful. They can sign, develop, and bankroll artists helping them get to market and sell records, which will give them a return on their investment. Naturally sometimes these best-laid plans do not come to fruition and the label loses out, but in many cases large artists can reap considerable profit for the label. We'll discuss this a little later when we look at the deals for the producer.

The world is changing. The landscape of the record label is something that has been developing constantly over time since its inception, but that rate of change has dramatically accelerated in recent years. Nevertheless, the record label continues to play a pivotal role to many artists and therefore producers and engineers.

The current shift in power we're experiencing with companies such as Live Nation, predominately a live event organizer signing big acts such as Madonna, opens up uncharted territory. We must note the fact that the world of music is changing and consequently it is very difficult to predict the future of how the industry will operate.

In this section we'll discuss the interactions you may have with labels in the coming years. On your career path as a producer, the record label will be a body you'll deal closely with from time to time as a source of work, or the conduit through which the music is released. This chapter will introduce you to the entity and who, as a producer, you should communicate with.

This introduction is not intended to be anything other than that, an introduction. We would recommend further reading (see Appendix F-1 for more information) should you wish to know the intricacies of a record label.

TYPES OF LABEL

Labels for many years provided not only the machinery to get a record to market physically, but also the means by which to do it: the studios, the producers, engineers, and so on. As we often refer to in this book, the Abbey Road model is one we can discuss readily. EMI in the U.K. owned the label, the studio, and on the whole employed the producers, engineers, tape operators, and assistants. In fact the label owned or employed most everything in the whole process of bringing a record to market.

As time went on, this changed and slowly the producer and many engineers became freelance, opting to be paid per job, or for royalty cuts on each record. Therefore these individuals would take an element of risk, just as the label had, with each and every artist. Those that make it pay dividends for all concerned, but those who regrettably do not make the charts and sell, produce less revenue in the longer run.

The labels were then challenged by the arrival of a number of independent labels that were not limited by constraints of large management teams, and therefore much of the music that could be signed and released was of a more underground quality. However, many of these labels were over time bought up by the majors and either stripped of their valuable acts or, in the case of larger indie labels, kept in their entirety (Harvest as part of EMI, for example).

This model remains more or less intact today. For this reason, the music production industry, and the role of the producer, is an unsettling one. It is therefore reassuring to see some producers combining high-profile artists on their client lists as well as experimenting by developing smaller, less successful, acts. In so doing, they are essentially acting as the label and are entering the market too.

In essence there are, at the time of writing, the famous 'big four' labels: Sony Music Entertainment, Universal Music Group, Warner Music Group, and EMI. If there were to be a fifth collective, it would be the scores of independent labels of varying levels now able to compete on a similar level playing field because of the Internet. The aforementioned Live Nation is now a serious contender, having signed a number of high-profile acts such as U2 and Madonna. Their case is different since they do not claim the copyright to the music recordings in the same way a traditional label would. We'll touch upon this in a little more detail in Chapter C-1, What's the Deal?

The independent labels are still a major force in the industry, bringing smaller signings to market. They can still influence underground movements and become large organizations in their own right. Look at the U.K. label WARP which represents the electronica market during the '90s through to the present day and has grown to become a large organization.

As a producer you might find yourself working with or for any of these labels and having a knowledge of the key personnel within them will be helpful. Below we provide insight into some of the key departments.

MANAGEMENT

The labels are naturally headed up by a CEO or similar role and there will no doubt be a board to which the senior management team are accountable. It is unlikely that you'll have too many day-to-day dealings with the management of the larger companies. Their role is naturally to maintain the business operations and some are better at engaging in the musical matters than others.

ARTIST & REPERTOIRE, AN INTRODUCTION

As we introduced in Chapter A-1, Quantifying It, the Artist and Repertoire (A&R) departments were once very powerful people within the record company. The A&R executive would offer new talent to the label, but would also in the early days produce the records in many cases. These people would steer the recording and sometimes sonic image of the record.

Consider George Martin, the head of the Parlaphone label, and the profound influence he had on the Beatles and their music. Martin was one of the pioneers who changed the status of the music producer; from being an employee, possibly with an additional role (such as head of the label) to a freelancer.

This department will be something of a link in a chain for your employment, connecting you with artists and the label. Understanding how they work and how the engagement happens will be of paramount importance.

A&R has "hugely changed" reports Joel Harrison, former head of A&R at Island. "It used to be about discovering music early, signing it before anyone else, and then building and developing them until they were ready for release. Now it is much more like Dragon's Den" (a popular TV show in the United Kingdom where commercial ideas are presented and critiqued by business leaders).

Harrison explains: "A&R are now asking of acts 'who have you supported on tour?' and 'which radio stations are playing your songs?' in addition to 'how much merchandise do you sell'." Harrison explains that the whole focus has shifted to proving the marketability of the artist and their buzz long before a major label will consider signing. A good example of this is Enter Shakari, who have gained a large following over many years and sooner or later success was going to follow, but not necessarily from a major label. The risk-taking and the investment that the majors used to take out to develop artists appears to have subsided.

The role of the A&R executive is much like a producer in so many ways in that there are many facets: the producer's role may range from musician to counselor, to financier, to manager, to master liaison officer, and so on. Meanwhile an A&R exec has a simple function as Harrison identifies: "to identify talent and then to make it work however you can." But the roles can be as wide-ranging as that of the producer: counselor, agent, artist development, stylist, press officer, financier, manager, and so on. Harrison surmises, "…at the end of the day, you can only be as good as your act. Full stop!"

The A&R team will identify an act they like and believe there's time to invest in them. This might be through many visits to gigs and hearing all their recordings to date. If this happens, the A&R executive might choose to become friendly with the act or their representatives through parties, meals, and so on. Meanwhile their music will be played at A&R meetings to gain consensus and aim to get the support of the heads of marketing, promotions, and A&R at the label. Says Joel Harrison: "…you want the heads of the areas to get emotionally involved in the music, invested even." From here the legal bits occur where the label will make negotiations and eventually sign the act while fending off any other interested parties. "Then a big dinner to celebrate," he says.

What do A&R scouts look for in producers? It's all to do with the style of the music and what the outcome (vision) is to be from the label. "Some you hire for their style. For example, a Stephen Street record sounds very British. Think Blur, The Smiths, and the Kaiser Chiefs. Whereas a Gill Norton record is more U.S. sounding (the Pixies, Foo Fighters, Jimmy Eat World)," explains Joel Harrison.

How do A&R select them? This is where the producer managers come in. Perhaps you might send demos out to them and "see who bites" continues Harrison. However, he acknowledges that if you're going for the cheap option, you could "go for their [the producer's] engineer or assistant."

Therefore, as a producer, you will have a fair amount of contact and discussion with the A&R departments, discussing acts and the label's vision for the act.

ADMINISTRATION (BUSINESS AFFAIRS DEPARTMENT)

As with any glamorous firm there will be people behind the scenes who engage in the administration and the business of the firm. Not everyone in TV is in front of the cameras, and there are people behind the scenes in the fashion industry too. The same is obviously the case in the music industry; behind every great act is a great set of administrators. Most likely the most connection a producer will have with these personnel is to chase invoices or to find out information relating to artists or the sales figures.

THE LEGAL DEPARTMENT

With any label deal will come a contract. As a producer you'll have a deal in place with the label describing what you'll do and what you'll get paid and when for the work you'll do. As we'll learn later, this is an important aspect to get right from the get-go with the label. You're unlikely to often liaise with the legal department unless you choose to represent yourself, as you're likely to use your lawyer, or your manager will handle this for you.

MARKETING AND PUBLICITY DEPARTMENTS

The marketing and publicity departments create the image, the marketing strategy, and promote the acts. They will have a handle on the images that support the artist's physical releases and will also be placed as promotion around the media through radio, television, and magazine interviews.

Product management

Within the marketing departments will be a product manager whose responsibility it is to liaise with all other departments to ensure a particular band or product is coordinated across the label to maximum effect. You may have some communication with the product manager depending on your involvement and interest in the artist's project.

OTHER DEPARTMENTS

There are many other departments which, as the producer, you may liaise with from time to time during your career. These departments are those that take your productions and work with the artist to promote and market the material.

New media department

This department's role is to oversee aspects such as the creation and promotion of the music video and perhaps the artist website. As the Internet becomes more powerful, new media departments begin to consider new innovative ways to promote music. You may be asked to contribute to this with interviews or take part in a behind-the-scenes film.

Sales and distribution

Sales and distribution may be a valuable department to you to ascertain how your production is selling and how you might gain an understanding of the success of a particular record, should you wish to.

Business support

Whatever anyone says, the minute you work for yourself, you're self-employed. What that essentially means is you're a business, as we'll discuss a lot more in the next chapter. You may choose to gather a number of people together to assist you in this regard, and these roles are scaleable dependent on the size of your activities. For example, you might prefer to work as a one-man-band going in and out of studios working with various acts and in this instance you may have a producer manager and a part-time accountant or bookkeeper. However, as you grow and take on three acts at once and a couple of engineers, you may need to scale up the accountancy day-to-day involvement.

ACCOUNTANT

Most likely an accountant will become a necessary part of your professional team as time goes on. As a relatively small outfit you can, of course, manage your finances accordingly and file your tax returns with some relatively sensible and diligent organization. Alternatively, you can employ the professional experience of an accountant specializing in the music and media sectors.

Employing a specialist accountant will, of course, be of benefit. They will be able to assist you in saving money through the tax system because they understand the allowances afforded to musicians and music professionals over any other regular career.

You will be able to seek out specialist accountants in your territory for what they can offer. There will be a wealth of these professionals available to you who will be able to assist you to manage your accounts accordingly.

LAWYER

You will, no doubt, from time to time need to call on the services of a lawyer for assisting in deal negotiations or simply for advice. There are specialist media lawyers who understand the industry position and the current rulings.

Seeking out advice can be tricky, but you may perhaps find a lawyer through your professional body in the first instance. These lawyers will usually give a little time free of charge to discuss your issues.

ROYALTY INVESTIGATION

At times you may need to enlist the support of some specialist professionals who will investigate royalties that are owed to you. There are some firms that specialize solely in expertly investigating the complex issues of royalties and where income streams might not be reported to the collection agencies in the first place, and therefore you're unable to earn from them.

WEBSITE DESIGNER

Many producers now have their own websites detailing their work and their credited albums. It is a relatively new shop window for the producer, offering them the ability to become more connected with an audience that may admire their work.

Many producers have chosen to develop their website themselves using iWeb or similar, but many will also consider employing a professional to develop and manage a site for them as they may be too busy to do it themselves. In either case, it would be sensible to organize your work on some form of site, whether self-managed or professionally developed.

As we mentioned in this book, other additional web outlets may include managing a MySpace page for you as a producer, and certainly the artists you are investing your personal time into will require some development in this area. The same should be developed, perhaps to a lesser extent, using Facebook in addition to the more professionally focused LinkedIn, which does house a number of music industry professional profiles, and may be of use in the coming years.

If you have enough administrative support, you may have someone manage your social media, or they may relieve you of enough mundane work that you're able to tweet about the more interesting things, such as the monster beat a hip hop artist has made in your session! All good PR for the forthcoming record!

PUBLIC RELATIONS

Public relations (PR) firms can come into the realm of the producer from time to time. The high profile producers will no doubt call on the expert services of a PR guru to assist in a campaign to hype up their new clothing line or even the new fragrance!

The PR firm may also need to be called on in other times, perhaps when a producer is trying to release an album on their own label, or heaven forbid, something in their lives is made public which they'd rather not have done in the first place or have simply kept quiet.

ADMIN SUPPORT

If you're in the studio constantly and wish to keep focused on the creative work at hand, but have yet to assign a manager to handle your work, you may wish to employ some help. In doing so, you may wish to consider some administrative assistance in some form; perhaps a PA? Someone to be there to answer the telephone, shielding calls for you and to organize some of the more mundane aspects of the role such as studio time, some session musicians, and equipment hire.

To some extent this will all depend on whether you're keen on managing these aspects yourself, or whether you're happy to remain creative within the studio for as long as possible. You will find your blend and tailor your assistance accordingly. There are no hard-and-fast rules.

Family and friends

A key part of the component in your success may be your family. Working within the studio business can be a very hard one to maintain while raising a family, and in addition it can be difficult to manage a relationship, or meet someone, when you're in the studio so much.

However, given the slightly fragmented nature of the producer's business these days with so little employment and more self-employment, family members are often playing a more critical role in the work of the producer. Many producers choose to work at home most of the time. Producer Tommy D works predominately out of his studio at home, in which he's able to develop artist's music in the comfort of his own home: the comforts that are opened up to his artists when they work, in many ways. While this is taking place, studio fees are lowered and it is possible to manage the process in a less rigid and formal way.

Given the nature of the home-grown producer, it is possible too that some of the administration might fall to the family at large. In some cases we've learnt that the producer's spouse has been known to manage some or all of the producer's work. Keep it in the family!

Networks and trade associations

As we mentioned in the last chapter, you may find that a rich source of contacts and business comes through your affiliations to a trade association or many of them. Joining these and taking an active role can be time-consuming, but you should find that what you get from your affiliation is neatly aligned to the amount you put in.

Belonging to these bodies can help you access networks you'd not normally be able to and get to speak to more established colleagues from whom you can learn a great deal. Even as a student, we would recommend becoming a member of one of the trade bodies and, where you can afford to, travel to take part in the networking and lecture sessions they may put on.

See Appendix F-1, The Tape Store for more details of some professional bodies in your area or how to search for them.

CHAPTER B-3
Being a Business

We're often amazed at some of our students. They believe being a music producer is all about glamorous studio sessions, parties, and gigs. But it's really a unique blend of business and creativity, regrettably not even in equal measure!

The world of music production relies on the flexible and independent services of a whole host of professionals. Over time, as we've already established, these professionals are of a freelance nature and manage their own business.

What this means in effect is that as soon as you work for yourself, you're actually a business. Not a large corporation of course, but still a business and one that needs to pay attention to the rules, regulations, and taxes that your territory insists upon.

In this chapter, we'll look at what it means to be in business. Being from the United Kingdom, we may speak with a U.K. focus, so you're advised to seek out the relevant support from your trade association and local business advisors for up-to-date and pertinent support as you set up shop. We've provided a list of many of the international trade associations toward the end of this chapter.

WHAT, I'M A BUSINESS?

One of the hardest things to learn when leaving school or university is that what you have learnt might not be the most important set of skills you'll need! We spent some time in the last chapter speaking about the art of communication and networking—important skills in an entertainment industry such as ours. However, some schools, colleges, and universities providing education in the music production sphere neglect one of the most important aspects: business!

By business we do not mean a mere unit or module on the music industry, although this is important as it sets the landscape and context in the career the student may work in. We mean business in the raw, industry-standard sense of the word. Things like good practice in administration such as invoices, remittance advice, communication, customer service, as well as those aspects of real life such as the business plan, forecasting, cashflow, bookkeeping, among many other boring, yet vital, aspects.

What is Music Production. DOI: 10.1016/B978-0-240-81126-0.00005-6

As a producer, unless employed by a label or production company, or you've hired a producer manager, you'll at some point need to have a handle on all those key business methods we've mentioned above. In this chapter, we do not begin to offer detailed advice on this topic, as we're not the experts. Instead, we offer a comprehensive overview of the things you'll need to consider as you set up in business and where you might wish to look for that much-needed advice.

As you begin on the path of the music professional, you'll wake up to note that if you really want to make a living from what you do, then you'll need to get the business side of things sorted. You'll need to consider a whole new set of issues, possibly outside of your normal realm of thought: music production!

Running a business can be challenging and demanding for every line of work, but in music production it can be harder in some ways. Regrettably, this is an aspect of the producer's work that many do not speak much about in the interviews you read. To many, once the "points" are out of the way, people are somewhat less interested in the goings-on behind the scenes, but are predominately interested in the goings-on in the studio and music itself. We agree, it is more interesting, but logic suggests that good housekeeping behind the scenes ensures more successful goings-on in the studio.

So "good housekeeping" could be a good title for this chapter perhaps. Much of this book deals with a discussion of what a music producer is and does, while this chapter discusses the aspect of simply being a business.

WHAT DO I NEED TO KNOW? THE BASICS

Being a business is pretty much the same for all lines of work. Regardless of how the money comes (profit is, after all, the ultimate reason for being in any business, unless you're what is known as a not-for-profit, or social enterprise), you have an obligation to work as a professional and stay aboveboard.

Setting yourself up as a business is therefore key to being aboveboard. In the U.K. there are many ways in which a company can be set up and it would be foolish for us to write a long and detailed example of these types here, given the international audience that may read this book. However, some aspects are important and relevant as we go through the discussion, but be sure to check how the points made relate to your country's laws and tax rules.

Deciding to go into business is never an easy decision to make, and often comes out of necessity. In the current economic climate, many new businesses you speak to say they were forced to make a go of it because they had lost their main line of work through cutbacks or so on. As a result, a number of businesses known as small or medium enterprises (SME) have sprouted up, which may indeed be you.

However, some business owners choose to start up their business in a calculated and completely prepared way, deciding to leave the house engineer post they had, and so on. In this way they can save, look for adequate finance and premises while all the time building up an appropriate client base as required.

Business types

SELF-EMPLOYED

As you kick a career off in the freelance world, it is likely you'll start out as self-employed. Anyone who is not working for a company is most likely to fall into this category. In the U.K., anyone who is a director of a limited company (Ltd) will need to return the same self-employed tax returns.

Being self-employed is about the most practical place to be. It's good as it's flexible, nimble, and allows you to move with the industry. You can work part-time for someone, while also being self-employed for the freelance work you get in. It is, in one sense, the simplest way of getting your tax returns completed with the least cost, as you do not need to employ an accountant to do this for you.

You don't have to register a company name such as with a limited company, although you'll need to let Her Majesty's Revenue and Customs know you're self-employed. However, instead of trading as your name, you might wish to consider a company name to develop a presence and brand, something we'll cover in a little while.

Being freelance can have its downsides also, insomuch as it is not a recognized limited company, which to some people is considered to be more safe or more established. This can be problematic if you're trying to get business from a larger corporation that insists on working with limited companies. This is unlikely to happen in the music industry, but might! You might also find, when looking for investment in your company, that the banks do not look as favorably on you as they might with a limited firm.

Nevertheless, this is the most likely place for many producers to be starting out and should be looked into in a little more detail through one of the sites below.

> For the U.K. see the Self Assessment website, *www.hmrc.gov.uk/sa/index.htm*
> For the U.S., navigate to *www.irs.gov/businesses/small/index.html*
> For other countries, seek out your government's pages.

LIMITED COMPANIES

The type of business and how you "incorporate" yourself is important. For example, you may wish to ensure that all your assets are protected should you run into difficulty because of a lawsuit. In the U.K., a limited company is a good option as it allows for some protection. The company is owned not by the individual solely but by either its members, shareholders, or the public. These three varieties make a distinction in that the company is separate from your own finances and your responsibility to the company.

In the U.K., the limited company can be one of three kinds:

1. Limited by Guarantee, meaning that the members of the firm are responsible to a limited amount in the case of the firm folding.
2. Limited by Shares, meaning that the shareholders hold the limited liability should the firm fold. This is the most common of limited companies for private firms.
3. Public Limited Company (PLC). This type of company is where the firm floats on the stock markets and its shares can be bought by any public individual.

You are the director who runs it. Should the company owe money through receivership, its assets will be sold to pay back the debt. (Assets are classed as anything that can be sold for cash, so in a music production Ltd company this could be all the studio equipment, for example.) In this eventuality, there are penalties for the directors (and these seem to be more and more financial in recent years) such as not being able to act as a director for a number of years once bankrupted. Essentially a limited company is run by the directors, governed by state regulation and shareholders are protected from some limited liabilities.

For more information in the U.K., see *www.companieshouse.gov.uk/*, a government organization with whom you need to register your limited company and file annual returns to.

PARTNERSHIPS

There is another limited company type, and this is perhaps more suitable for the music industry professional wishing to work in a partnership. This is known as a Limited Liability Partnership (LLP). It is similar to a limited company, but does not require a larger board of directors. It can simply be two partners working on something, sharing some of the benefits of the limited liabilities enjoyed by limited companies.

Partnerships can also operate without limited liability. Essentially each partner will have to produce self-employed accounts, in addition to partnership accounts. A little strange, but many people work this way in the U.K.

Finally, without being an employee of someone, you have the option of being simply self-employed. This is where you manage your earnings and declare the tax you pay each year by submitting accounts to the U.K.'s Inland Revenue. (A freelance sound engineer, for example.)

THE U.S

In the United States, the terms vary slightly in that there are two equivalents of the U.K.'s limited company:

1. Limited Liability Company (LLC)
2. Limited Liability Partnership (LLP)

The first is akin to the U.K.'s limited company, and the Limited Liability Partnership is broadly the same as the U.K. equivalent.

There are other company types in the U.S., such as corporations (also referred to as a C Corporation) which are larger companies where shareholders have placed money into the company in return for their shares. An S Corporation is slightly different in that the tax affairs of the company fall to the individual shareholders in their own personal accounts.

For more information about these, seek out further information at *www.irs.gov/businesses/index.html*

BUSINESS IDENTITY AND BRAND

That's some of the boring stuff out of the way, although very important nevertheless. Onto the more interesting and engaging things for some people: the brand.

The business identity should represent what you're about. Are you setting up a production company with premises and so on, and are you focusing on one specific genre? If so, you may be able to tailor the whole package to appeal to the genre and the people within it. You'll need to choose a name carefully and design a brand that exudes your position in the market, values, and aspirations for the future.

Business name

Every company has a logo these days. Every company has a name at least. Choosing these can be important. Some companies are simply named after someone and, given time, in the market the other manufacturers' products are not called vacuum cleaners, but Hoovers, whether they're made by Hoover or not! The same goes for public address systems; many people refer to them as the Tannoy but in fact this is the name of a specific loudspeaker manufacturer who hasn't necessarily produced the system being referred to at all. Other companies are named specifically to fit into the world in which they trade and their activities. For example, The Digital Audio Company in the U.K. pretty much says it on the tin, as does Audio Recording Unlimited in Chicago! Actually, if you delve a little deeper, both of these firms are not simply recording studios, as it happens.

Choosing the right name for you will depend on so many things we'll not delve into here. You may choose to call it Your Name Productions Limited, or something completely different. That is up to you, but whatever you decide, be sure you're happy with it and that it will stand the test of time. One thing to consider is choosing a name that does not restrict the activity you do. For example, if you were to call it Your Name Vinyl LP Limited and you've become in time a mastering house dealing exclusively in Digital CD and MP3 mastering, the name no longer suits. Make it suitably vague to the future possibilities you'll want to encounter.

URL

Before spending a lot of time developing your brand and image, make sure that you're the only company with the name you're looking to be and if you are, ensure that you get the URL! There's little point going out and getting a new brand for yourself to later find that the .com, .co.uk and .co have all gone, unless you're happy with .net or .org.

Go online to one of the registration companies to see if the name is available and, if it is, why not register it? Should you change your mind later, you'll have only wasted at most $20! Having this is important as it is your calling card and will be a large trading place, or certainly somewhere for your future clients or fans to pop along to find out how to connect with you. It's all about the network!

Business logo

Every company these days will have a logo or image. This is important given the number of images we need these days for the website and social media, in addition to the normal time-honored traditional signs, letterheads, compliment slips, and business cards.

Choosing a logo is something that you can go out and do yourself, but you might choose to get a professional to do it, as it may have more impact and a longer lasting appeal than something you think you like. The image and there-fore the business logo can become the brand in its own right, so it is worth spending a little time investing to get this right. It is important to remember that music production is a creative pursuit and therefore you might want this to be reflected in your logo. For example, if you are a hip hop producer, then you don't really want a logo that looks too formal.

Stationery

Once the logo is decided on and the image is coming together, it's time to get a set of business cards printed, as well as the business-standard compliments slips (if you're in the U.K.), letterhead, and so on. These will be the image your business exudes as it does its business. It might just be Your Name Productions, but to have a card to give out to the right people will be very important.

Thinking about the business cards for the moment, to have something strik-ing and different from the norm can help in a big stack. Coming back from a trade show, which the average producer might wish to attend from time to time, you'll have a whole stack of cards which will remain there for a while (unless you've read the time management and organization bits later in this chapter).

Even if a contact has 50 cards from the latest AES convention, if he can gravi-tate to your card because it's unique, then he'll find it interesting and he is more likely to get in touch. One card we were given recently was unique. It was quite thick on exquisite white card, almost reflective a little. It had no writing on it whatsoever. All the text was embossed on the card in such a way it was easy to read. The company also took the advantage of placing the details in Braille too. Very thoughtful!

Another card to the eye looked the same as most others, but this one opened out to be a little booklet. This offered a lot more space for the company to write about what they did and how they helped their clients.

Either of these tricks can be good and will make you stand out and be remembered the morning after the night before, and in business this counts. So think of a way of making a business card that's interesting. Stick with the regular size, however; people don't appreciate a card that does not fit in their card carriers or wallets. You might choose to get the cards printed with a gold or fluorescent yellow around the edges, making it stand out. Have a think and get some advice from your local printer or designer.

Another trick some business gurus recommend is to include space on the card for you or others to write some notes on. For example, many networkers suggest that to write some pertinent information about the person on their card will assist you in remembering them both visually and who they are. Leaving some space on your cards for others to write on them would also be helpful.

The remaining stationery is fairly straightforward and should reflect the business card, unless you're the business using the embossed card. The design of these should be professional in style and make an impression. It is a good idea to get these professionally printed on good paper, as this beats hands down most printing facilities available to you as an individual.

Website

You'll probably choose to have a website, unless of course you're just a sole trader and are always in the studio in employment. You may not see the need to have your own site. However, you might feel the need to connect with your audience and colleagues. As we'll cover later in the book, you can manage this yourself. Anyone with a Mac and iWeb can pretty much sort out a fairly swish site, using the images provided from your logo designer. The content is then up to you. There are of course similar WYSIWYG (what you see is what you get) packages for the PC too.

Alternatively, you can employ the services of a web designer to create something a little more exciting for you. This might suit if you want the website to be more interactive and fun to visit. Alternatively, it might be wise to employ a designer as you might be too busy or not have the patience to understand domains, hosting, and so on.

Either way, a website will be good for you and, along with your URL, it will become a key identifier. Your email will belong to the URL and as such will maintain a level of professionalism and belonging to the firm.

What you place on your site will be up to you, but most producers will place their discography and perhaps some photographs of them in action in the studio. Placing the services you provide on the site would be beneficial, especially if you have a specific skill or specialism, such as a specific genre.

BUSINESS PLANNING

Read any business start-up manual and the term "business plan" will crop up in more than one place. Another dull, yet important, process to go through. It can be vital to map out how you wish the business to develop and expand. Business plans are, if you like, a roadmap for the future of the company, detailing both ambition and the real-life fact of how the business might achieve its heady goals. However, realistically, for music production start-ups it might not be all that sexy or indeed all that necessary. It all depends on how big you might want your business to become.

Business planning in many forms of business can be straightforward. You devise a business, write an elaborate plan with all the correct facts and figures you have to hand at the time, and then perhaps tout around for some funding from a bank or venture capitalist, if that's what you need.

We make it sound easy, but it can be far from it. These plans often need to be detailed, with profit and loss analysis and some assurance that the business will be a success. Nevertheless, if you're wishing to translate your ideas into a larger business with investors or fellow shareholders, you'll need a plan.

Bringing a music production business plan to a bank could be problematic. The financial returns are often sporadic and cannot be guaranteed. Gaining finance for such endeavors from a bank can be difficult and therefore music professionals frequently take a more organic approach to building their businesses, with often family loans or simply chipping away, gradually amassing their own financial leverage.

PRINCE'S TRUST

One organization worth mentioning is the Prince's Trust. (Unfortunately, this is only available in the U.K. although there may be similar funds available in your country.) The Prince's Trust is predominantly set up to assist and support disadvantaged young people who are not in work, education, or training. However, even though you may not fall into this bracket, the organization may still be able and willing to support you with start-up business loans if you have a good idea and plan for a business but are unable to gain the financial means to do so. It gives both practical and financial support and to creative industries. See the website for further details. *www.princes-trust.org.uk*

Producing a business plan need not be a chore and many banks provide business planning forms, software, or online tools to help you along should you feel the necessity. It is worth getting some of these packs before you set up your company. It's worth it, for it not only helps you shape your business (unless it's clear you're a self-employed producer and that's all you're up to), but also will give you a clear indication of the bank's offer as you may choose to create a business account.

What's in a plan?

Every business plan will be different, but it will contain mostly similar content. As we've already established, you may choose to do a plan based on the fact that you want to plan the business out clearly as it goes forward, identifying key goals you'd wish to achieve in time. Alternatively, you may be pulling together a plan to go to someone for some key investment, whether that be a bank or investors.

We'll take a quick look at the kinds of things you'll need to think about and develop as part of a typical business plan. These are not exhaustive explanations and it is advised that you seek out books solely on this subject matter before completing the full business plan.

You'll be expected to come out with a business summary that details the business, its purpose and its values. You might also wish to explain where you want the business to be and grow into over a three- or five-year period.

Next you may give an overview of the business and how it operates. Remember that many people do not have the knowledge of your industry that you do and will need simple and positive explanations of the work you do. You will be expected to lay out the financial position of the company at this current time.

So, in this example you may be a self-employed person who has produced several albums and now wishes to invest in a studio complex to expand the number of albums you can work on and also claim some of the studio charges. In doing so you will need to explain your current income streams and how the new business will change and increase your productivity. This is known as the business strategy.

By providing a business strategy, you outline the ways in which you are confident that the work will come your way and the methods by which your company can honor the new work. It is here you can expand on your values and expectations for business growth. Alongside this, you will need to establish the marketing strategy if you need one. Alternatively, you will need to outline how you intend to maintain the level of work coming in or attract more business.

As you grow, you will no doubt take on staff to support your studio complex and the increased work you've taken on. It is worth using the business plan to outline how you might envisage this working. Will you have an assistant, a Pro Tools operator, a receptionist, and so on?

Finally, you will need to submit some financial forecasts and budgets. These will accurately demonstrate to any investor that your business to date is strong and that you've considered the future and the long-term financial risk you're taking. Running a studio complex, or any business for that matter, may not be profitable for the first three or so years of incorporation. Earning enough to start to pay back the bank can take that long a time and is not too uncommon.

Before completing and finalizing your business plan, it will be worth you seeking out a local business advisor to assist you. Their experience might not be in the music industry, but they will know what a bank or investor might be looking for in your plan. Just like a producer is to an artist, or a mastering engineer is to a mixer, the business advisor will provide an objective viewpoint.

DAY-TO-DAY WORK

The day-to-day work of the producer is much like any industry and the better you are at it, the better you may be in business. Again, this stuff is not the magical creative music-making that flows in the studio from time to time, but is still vital to have an understanding of.

In this part of the chapter we'll look at some of the more mundane, yet vital, skills and decisions you'll need to make as you go through your career. This chapter will probably be the least read in the whole book, but for those of you who are students of music production and wishing to go into the industry, we hope you'll gain a lot from this now through your studies, and throughout your career.

Producer manager or not?

As you read in Chapter B-2, Your People a producer manager can be an important person in the whole process of running your business. As we learnt from Joe D'Ambrosio, a producer manager might take care of so much of the work you do on a day-to-day basis that some of the items below may fade from focus. However, there are some gems in there, such as time management, for all of us.

Whether you wish to engage a producer manager or not is entirely up to you, and as you get busier it might be essential for you to have representation to go and get you work, and continue to get you work. Earlier in this book, Joe D'Ambrosio spoke of the work Jay Newland continues to get. Newland has been "constantly busy for five years" since the release of Norah Jones' first album. Keeping this flow of work is what you pay a percentage to the producer manager for.

The decision as to whether you choose to hire one or not will be based on the work they can promise to provide you and also the added services they can offer. If you'd like your manager to take care of all the financials and any other aspects in the business you'd rather not deal with, then they will become an invaluable member of your team.

Time management a.k.a. productivity

Time management sounds almost cult-like. To many people, they follow their system religiously. In fairness, this is how positive gains manifest themselves. Many producers might read this section and laugh out loud and consider it barmy! They may be right too! Once they've calmed down you'll find they have their own system and they use it religiously.

Effectiveness can be realized through the productivity system chosen. There are many ways this can be achieved. You'll note the word *productivity* used here in place of the more standard *time management*. There is a reason for this. Time management only looks at one aspect of being effective. A producer's time can be managed, but it's more about doing, planning, and organizing than time alone.

Setting definitive times for things to get done will stifle creativity in the studio, but not outside for other, more mundane tasks. So time is not the main factor here. Productivity is! If you think this is all common sense, it probably is. But we bet you could do with a refresher and admit that there is still room for improvement.

DOING

This is about finding the time to do things. As a producer, you might have a lot of time where you're in the studio and this will not always be necessarily 9 am till 5 pm. The mad days of sessions lasting for days might be over, thankfully, for the majority of sessions, but it is not uncommon to start around 11 am. What do you do with that time in the morning?

If you become regimented with your routine and organize yourself accordingly, you could be up, out of bed early at a set time every morning and use the time before the studio to think, respond to emails, update your social media and website, and do some much-needed administration. This routine will enable you to be ahead of the curve and very much in control of operations.

Finding a rhythm that suits your way of working enables you to have the time to control the flow of work coming your way and will make you productive. It will allow you to achieve those things you're trying to achieve. Take a look at your schedule and think about making time to process all that administration and give yourself some time to think about bigger things.

PLANNING

Once you have found some time, you need to consider the planning of a few important things. As we'll cover later in this book (Section C-3, Project Management), there's a whole set of project management tools for organizing large projects such as the recording of an album. This planning, and your own productivity, is all about what you're doing, who the clients are you're working for, and who you might like to start working with in the future.

Providing the space to consider the large stuff like the direction or path your career is going will assist you in making smart and sensible choices as options unfold. However, this time is also for you to consider the smaller things you need to plan in your life such as when to take a trip, hook up that meeting, or spend time networking on the phone or in person.

As for the recording projects you're involved in, you can use this time for that too. We'll discuss project management later in the book (chapter C-3).

ORGANIZING

You've carved yourself out some time to consider what to do and when; then, of course, there's the point of actually planning it and organizing all those things we want to do.

Finding a system that effectively ensures we deliver on everything will be a very personal thing. Some productivity methodologies have become really popular in business circles. Cult-like, in fact. In the past few years, systems such as David Allen's Getting Things Done® (also known simply as GTD®; see *www.davidco. com*) and Mark Forster's Do It Tomorrow (DiT) (see *www.markforster.net*) have become widely adopted. Longer-standing systems have been developed, such as the *Seven Habits of Highly Effective People* by Stephen Covey (Free Press, 1989) and through which a whole other planning and management system is created and followed passionately (see *www.franklincovey.com*).

In the following section, as you laugh aloud, we'll look at how such systems relate to anyone in business, and you might even find them helpful as you manage multiple projects, artists, and events as you get busier.

"I've been so busy lately… my diary is my saving grace."

Tommy D

Productivity systems

Getting Things Done® is a system by which the user organizes her tasks by project and concurrently also by context. This is exceptionally powerful to people of all walks of life. The project might be the production you're working on. There might be other projects which are either in the planning phase or on hold, and these might be contained in their own little projects. Allen's contexts relate to the place or situation you need to be in to complete these actions. For example, studio, office, home studio, mastering studio, rehearsal studio, record label offices, meeting with manager, meeting with an artist, and so on.

The premise is that, through the use of software or sensible organization on paper, one can quickly identify the tasks that need to be carried out in the home studio while you're waiting for that DVD to burn. Also, you can then decide what to do when in front of the computer stuck on a train going to a meeting. Getting Things Done® can work very well, but relies on complete dedication to the art and also a fairly rigid system.

Many find the system fantastic and develop the skills and tools to reflect their own work or merge it with other systems they have researched. Others report spending so much time playing with the systems, they fail to become more effective!

These systems have been developed by their creators to assist their adopters in handling the sheer amount of things professionals are expected to do these days in busy lives. They are in no way music industry related but can be adapted accordingly.

There is no reason why Getting Things Done®[1] or Do It Tomorrow could not be applied to the life of a record producer, just like any other system.

This all might sound a little far-fetched. To many people, they follow their system religiously. In fairness, this is how the results manifest themselves: by relying on their system. Given the many people who have adopted these systems, it clearly must work. Many producers might read this section and laugh out loud and consider it barmy.

Try to stop them laughing for a second. Calm them down, and ask them how they manage their professional lives. As and when the laughter abates, you'd soon find they have developed their own personal productivity system to cope with the immense task of producing a record, or records.

For example, Mat Martin, a producer, musician, and artist in his own right (*www.kirstymcgee.com, www.hobopop.com*), employs his iCal diary, two spreadsheets, and his email to manage his day-to-day recording projects. Many producers may like to simply use the iPhone and manage accordingly.

"I run two spreadsheets—one for the money and another for the songs. Money is obvious. Songs is a record of who's playing on what, which day we hope to record, whether we have chord sheets done for them or not, et cetera. I update them both constantly," says Mat Martin.

This is a clear example of a system that has been produced to serve the purpose. Using spreadsheets works for Martin, but for others just good old pen and paper might be the way forward. Whatever the system, you'll need to be effective, as your time is your money and making every bit count will help.

Taking time to develop these systems and skills in the down time between productions should ensure that, when the times are stressed with juggling plates, you're equipped with a solution to overcome the workload while feeling free to engage with the production in front of you.

Administration

These curious productivity systems are not just about the time management, but encourage you to consider everything that you use for work. Filing crops up quite a bit for example, and suggestions are made as to why you should adopt a system that allows you to clear the decks and keep clear focus on the production you're engaged in.

Knowing that you need to be organized and do your administration is common sense, we admit, but we all have difficulty keeping up and completing it, don't we? We hope here we're able to encourage you to think differently about it and

---

[1]Focal Press and the authors of this work are not licensed, certified, approved, or endorsed by or otherwise affiliated with David Allen or the David Allen Company, which is the creator of the Getting Things Done® system for personal productivity. GTD® and Getting Things Done® are registered trademarks of the David Allen Company. For more information on the David Allen Company's products, please visit their website: *www.davidco.com*.

take a little time to get your skills in this area nailed when you have the time, because when you're busy, you'll want to get on with the music at hand.

Administration is naturally important. Having an understanding about what needs to be done and when will help the money come in, especially if you're ensuring you're getting the invoices out the minute you've done the work. However, making time to respond to your emails and doing so in a fairly timely manner will be very important too.

It's all about getting into a routine, or putting a system in place, relying on your chosen *personal information manager* (PIM) and note-taking system. Become an expert at it in addition to developing a conscientious attitude to your work and you'll do well. Consider also employing a database or *client record management* system to help you with your productivity, as we mentioned in Chapter B-1, Being a Producer.

Reading books such as the ones we've mentioned above will help you hone your skills in these areas and ensure that you've considered the way in which you could work, not just in a simplistic "how-to" way, but in a more holistic way. They encourage you to consider why you are doing the work you are doing, and where you want to be in the future.

Imagine being able to confidently manage the work you're engaged in, meeting all your deadlines while remaining responsive and proactive—not just fire fighting! Thinking about your time and the wise use of it will lead to more opportunity and hopefully success.

Note-taking

Believe it or not, note-taking is still an important part of working as a music production professional. As we'll learn later in Chapter C-1, What's the Deal, note-taking can be very important to ensure you capture all the information of a session. Noting such things as who played what and how, what you agreed on with the A&R person and the band, what order the tracks should be as you're recording them, the mic position on the bass drum, the settings of the Vox AC30 you managed to squeeze that awesome solo from the guitarist from, will help you keep track of what you need to remember.

Notes such as these may be extremely useful as the project goes on. Many people will manage easily to retain the knowledge required to recall the information. Others will get swept away with events and perhaps forget the odd piece of vital information. It is the retention that is the issue. There are so many aspects of the session you may wish to recall later on for historical or nostalgic purposes.

How you take notes will be completely up to you. We appreciate that many of us in music production are quite keen on gadgets and may choose to rely on our smartphone, but many would still suggest that pen and paper is the fastest and most creative for getting things down. However, some choose to use their laptops or iPads, such as Mat Martin mentioned above.

The most important reason to keep some paper and pen about is for any legal stuff that might be discussed. It would be worth taking anything that is verbally agreed down for reference. You may not be able to enforce any such agreements, but they will provide you with the notes to follow up on anything at the earliest point, plus allow you a record should you wish to revisit what you *thought* was agreed.

Accounting and bookkeeping

Yes, it can be boring. Being a business, and certainly self-employment, will mean that a lot of the day-to-day bookkeeping, and so on, may fall to you. It's not rocket science and it is not akin to programming System Exclusive in MIDI, or working out the detailed acoustic solutions to your control room. However, it does take a little organization and diligence to do easily. We'll cover the accounting bit later on, but here we'll cover why it's important to keep on top of things.

Getting used to collecting and gathering what you need to do to file your tax returns takes a little while to grasp. It is perhaps simpler now than it has ever been, especially in the U.K. with a comprehensive online tax return system. What you need is to get organized and manage your papers accordingly, and perhaps even open a separate business bank account to separate your business money from your own, as this can make the accounting side of things a little easier.

The country in which you live will have its own peculiar systems and regulations and you should, if you're not already getting on with it, look on your government's websites for further information. Here in the U.K. we're expected to report a number of details about what we have earned and what we've had to pay out in order to run our business. The online system (of the HM Revenue and Customs people) will calculate the tax you owe based on the figures you give. However, it is very wise to read up on the way in which this works in advance, as this might inform your purchasing for the coming year.

Being self-employed means that you have to purchase your own equipment and tools of the trade in order to provide your services. As such, these tools are tax deductible. Each territory will have its own system by which this works, but many tax regimes can be very generous when you purchase new essential equipment, providing tax breaks, and so on. Keeping abreast of the new provisions for this, and perhaps any other tax breaks on offer, may make new ventures or activities possible. So check that site from time to time.

Taking an interest in your business affairs will make life less exciting perhaps, but may allow you to make more informed decisions about the market and the future of your company.

YOUR WIDER BUSINESS AND OPPORTUNITIES

In this day and age, as we find ourselves in a global financial downturn, professionals in some sectors are finding their share of the action dwindling. The same is the case in the music industry and many producers are either having to spread their activities a little or are choosing to do so to add new income streams or variety to their work. The portfolio career is becoming widespread and is a method to insure financial security going forward. The reduction of all eggs in one basket is perhaps a wise strategy.

Along these lines, it is not uncommon for producers and other music industry professionals to find new ways of expanding their reach. Activities might include starting a band instead of leading other people's sessions (this is more likely for personal reasons, such as enjoyment, as in the case of Tommy D's band Grafitti 6). However, there is a pertinent reason for doing this as you become the songwriter, not just the producer, and as such your share of the finances is increased.

Other examples might be writing a book such as this or offering your skills up to education by means of guest lectures or master classes, such as those provided by JAMES (Joint Audio Media Education Services), question and answer sessions at stores, product endorsement as in many of the trade magazines, acting as a consultant in studio practice or studio ergonomics and design engineering, even some mastering. There are so many activities you can spread to, especially if you've made a name for yourself in production already.

Generating new business is never easy at the best of times, but successful producers have a track record and this can help their new ventures in many ways. For example, Phil Harding (producer of many hits, such as East 17's success in the '90s) has written a book about his time working at PWL and recording the likes of Kylie Minogue and Rick Astley. Other producers have spent time considering their work and placing this information in books, but there are nowhere near enough and we hope that many producers will spend time to reflect on their expertise and share their experiences in the fullness of time. Phil Harding is an example of a producer who has chosen to also enter both the education field, offering his time to teach students and give guest lectures, in addition to working professionally as a musician recording his own album recently. He has chosen to do both these projects without the backing of labels or publishers.

Professional affiliations and networks

We've mentioned the benefits of joining professional bodies before. Many of these are listed in Appendix F-1, The Tape Store.

Joining a professional body can be very useful to make new contacts and network. Joining an organization such as the Music Producer's Guild (MPG) or the Association of Professional Recording Services (APRS) in the U.K. or an equivalent in the U.S (AES) etc. can provide many opportunities to meet and interact with fellow professionals. Also it is an opportunity to discuss pertinent issues and perhaps raise awareness of the industry at the current time.

As with any professional affiliation, you get out what you put in, and there may be opportunities for you to join the organization and help along the way, opening more opportunities to network.

Similarly, there are many online networks such as Gearslutz and ProSoundWeb which provide professionals with connections with many other audio professionals across the globe. These can be a vital source of audio information and tech support should you need it. Take a little time to check these out if you can.

SECTION C
Prepping It

What's the Deal?

In Section B we discussed in some depth the idea of becoming a business and formally setting up a company or becoming a sole trader/freelancer and the various points you should consider and be aware of.

Our intention for this chapter is to raise your awareness of some of the deals that are made. We introduce some pertinent organizations you should be aware of and discuss, as an overview, the deals that most often occur.

We do not claim to be experts in this field and do not intend this chapter to be an in-depth or authoritative reference, as there are many books and websites out there dedicated solely to the legalities in the business, some of which are cited in Appendix F-1, The Tape Store at the end of this book. As a caveat, we would always recommend that you engage the expertise of a music industry lawyer if you are ever uncertain, and perhaps even if you are certain.

Although most people within the music industry don't just "do it for the money," it is obviously what keeps the music world turning and certainly you need to be able to earn a living from your work should you choose to fully immerse yourself in this career. As a producer, understanding what your slice of the pie could or should be and how you will receive it is no doubt of definite interest.

DEALS, AGREEMENTS AND CONTRACTS

We've all heard of those incredibly large deals that musicians sign to a label for. Many have indeed gone on to earn every single penny of it through good sales, making them a surefire investment for the labels again in the future.

For top artists this bandwagon, to some extent, can still exist. However, many artists have been gambled on and have signed such deals only to later not recoup the money banked on them by the labels. In the modern age, deals have adapted to reduce risk for the labels, and additionally to find new income streams for the investors in light of a reduction in record sales.

Deals and their associated agreement, or contract, are what make the industry go round. It is important to be clear from the start what everyone is entitled to. Making a deal and agreeing on it is not always difficult, but many do not do it as

What is Music Production. DOI: 10.1016/B978-0-240-81126-0.00006-8

efficiently as they should. For some, making a deal seems formal and rather too business like, as they prefer to concentrate on the music at hand and perhaps go with the flow. To delay may seem fine, but could be unwise as these things are better negotiated as early as possible in the process.

The deal is the principle behind the working relationship, outlining what each party will get out of the album or activity. This will need to be agreed and contracted at some point either in a formal contract or by some form of written agreements (see Key Advice for Deal Making sidebar).

KEY ADVICE FOR DEAL MAKING

The deal you make is going to be different nowadays. The income from the sale of music is no longer just the sale of the physical product. There are now the digital routes to market, whether that be a cloud-based service such as Spotify or from a digital service such as iTunes. There is the use of the recordings in films or television and even computer games. The industry and possible use of any music can be larger than the obvious CD sales.

While the routes to market are more spider like these days rather than a straight line, so are the deals and the clauses one needs to consider. It is important to be prepared and to consider your position when making all deals. Here is some generic advice for you to get started with and consider.

1. Understand what you're doing

Whatever you verbally agree to, evidence it in writing. Bind it! It can be the back of an envelope and as long as it is signed and adequately reflects the deal you've made, then that ought to be ok. It does not need to be formally presented.

Contracts are funny and fiddly things. Fiddly insomuch as they can turn friends into enemies if things go wrong down the line. What you agree in the studio verbally is also technically binding, but can be difficult to evidence in law. (See the section on note-taking in Chapter B-3, Being A Business for more information.)

So try to get everything in writing and remember it does not need to be a formal contract. Sometimes people will not be happy to sign things unless they absolutely have to, so best is to follow up with an email either confirming the agreement or encouraging a response saying something along the lines of the following:

> "It was good to meet you at the studio today. Great room to work in and fantastic material. I look forward to making a start producing the album on January 5, 2011. Shall we start at 10 AM? It will be tight that we have to record and finish the whole album by the end of February for delivery to the label, but we should work on the pre-production as discussed as early as possible.

> "I am pleased we were able to speak about the deal and that we agreed X, Y and Z with regard to the royalties. I trust my recollection is correct? I understand we have a restricted budget to work with and as agreed I'd like X paid up-front as part of the deal.

> My fee we agreed for this is $10,000 and I'll get a credit on the album as sole producer…"

The ideal is to get all the parameters of the agreement in writing about where, when, to do what, within what time period, within the specified budget, how the payments will work, who will pay me, and how the royalties are to be calculated.

2. Understand the copyrights involved

Copyright is the key to many income streams in music and as we've mentioned, producers will need to know now more than ever the different activities that can be protected. We discuss these in the Collection Agencies sidebar later in this chapter. Ensure you understand how the mechanisms work, as this will help you choose your actions and negotiations.

3. Confirm your entitlements

Be clear with the client that if you play something on their record, you'll want a performance fee and slice of the performance royalty. If you write some of the music with the artist, again you will need to agree in one way or another that you wrote it too and that it will be registered with the appropriate collection society. Whether you get a credit or not, you should ensure your work is in this way reflected in a one-off fee or similar. Again, get it all in writing.

INCOME STREAMS

As we have acknowledged, the industry has changed and morphed over the years. But how has it changed from the perspective of income and employment? We've seen models such as that of our traditional producer working for a large corporation, and the producers and engineering staff working under full-time employment. Concurrently during this period there were mavericks such as Joe Meek who was in some ways ahead of the game. Meek owned his own studio, his own label Triumph, and later his production company Meeksville Sound. Meek engaged in considerable technological and production innovation but also the business side of things and was a fantastic example of an independent.

Over time many producers began to shift toward freelancing and utilizing income streams in a different way. While the label might receive a considerable amount of income, the producer moved toward percentage deals of the album's profit. The opportunity to negotiate a fixed advance to get on with the project would be welcome in one hand while the other could enjoy points on the album's royalties (if successful). The latter, to some extent, appears a little like modern performance related pay.

Making money in the music industry can be difficult. As we'll discuss in a little more detail later, freelance producers, if just producing and depending on the deal, are likely to draw an advance from the artist's label. In addition to this, a producer may be entitled to a percentage (points) of the royalties, which is dependent on the deal negotiated.

Strictly speaking, if the songs were penned by a professional writer and passed to a nonwriting artist to perform, then the songwriter would gain their income

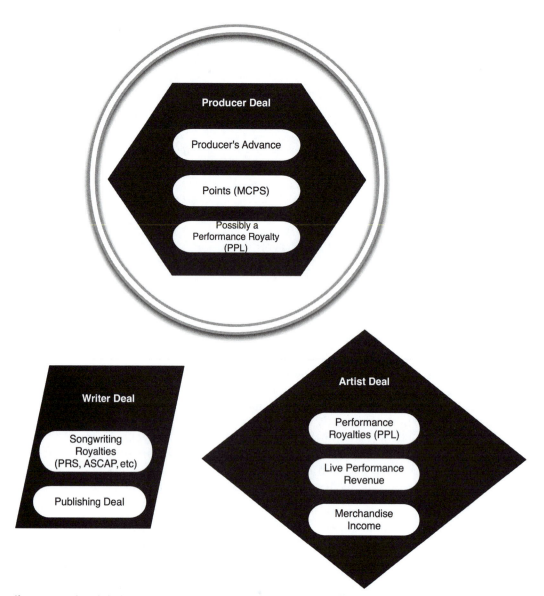

If we were to pigeonhole the income streams available to the producer, they'd just fall within the circle, usually containing deals made with the record company, which would include future royalty points. Similarly, if we were to pigeonhole the income streams for the songwriter (not performer), then these would be from songwriting royalties and perhaps a publishing deal. The nonwriting performer might only receive performance royalties and incomes from live gigs and merchandise.

from royalties each time the song is "recorded onto any format and distributed to the public, performed, or played in public, broadcast, or made publicly available online" (PRS for Music; see PRS section later in this chapter.)

Similarly, an artist who did not engage in the writing of the material they've recorded would be entitled to the performance royalties, live merchandise, and a share of the live performance sales.

In the real world, of course, these rather purist lines are blurred and some producers will write with their artists and as such will be entitled to a share of the writer's royalties.

Toward the end of this chapter are brief descriptions of a few of the main organizations (in the U.K. and U.S.) that take on the responsibility for making sure that you (the producer), the artist, and others involved in the process

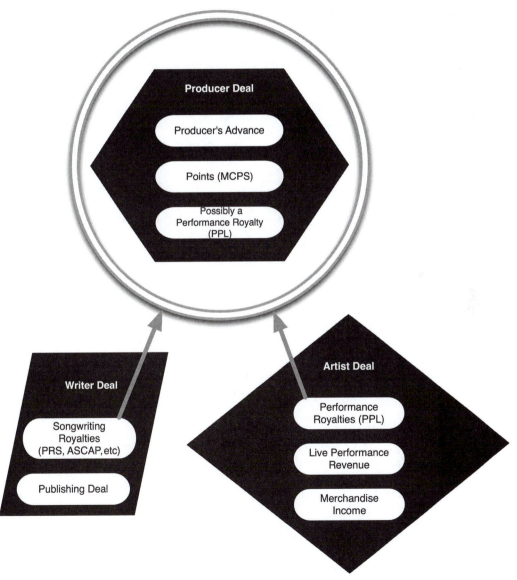

Of course in many cases the producer will be involved in much more than simply the production, perhaps performing on guitar on the track as well as cowriting it with the artist. In this instance other income starts to stream to the producer.

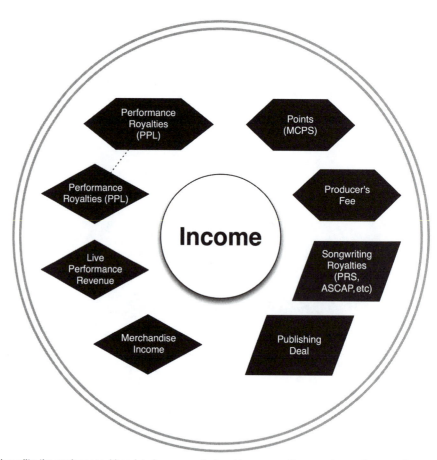

In reality, the producer could seek to become part of a wider source of income streams by expanding to become involved in all aspects of an artist's activity into what is known as a 360° deal.

of making a record get paid. You will have to register with all of these organizations in order for them to collect the money due to you, but this is a small price to pay in order to ensure that it ends up in your account!

Royalties

The big issue to consider when discussing any type of deal is the percentage of royalties that the producer may or may not be given. On a "costs" deal, as described later, the producer may receive a lower percentage of royalties due to the fact that they have received more money up front. This may be quite prudent if album sales are not expected to be that high.

It may be that a producer's fee will offer a greater amount of money than future album sales will yield. This is an unknown to a certain extent and if a record label intends to market and promote an artist's album heavily, some producers may forgo larger production fees or advances altogether in favor of a bigger royalty payment from potential future album sales.

As with all things financial in the music business it is the power of negotiation (ideally via lawyers) and the professional standing of the individuals concerned (the relative clout) that determine the finer detail and financial specifics of individual deals.

1 POINT = 1%?

When discussing deals, agreements, and royalties the word *points* is invariably used. A producer may be given a point on an album but the percentage cut, or share this actually relates to, can vary from deal to deal. The producer's royalties are usually taken out of the artist's royalty share (the artist in effect pays the producer) which obviously varies in itself, meaning that what the producer will receive in royalty income will also be dependent on the specifics of the artist's deal. Labels, producers, lawyers, and so on, now tend to talk in royalty *percentage* and therefore points (although a commonly used term in the industry) can be somewhat interchangeable nowadays.

COLLECTION AGENCIES

The collection agencies around the globe are incredibly important in ensuring musicians and music production professionals receive performance and other royalties for their efforts. We have taken some time out to speak primarily of the U.K. agencies here, with a mention of those based in the U.S. Website URLs are listed in Appendix F-1, Tape Store which additionally includes agencies from around the world.

The U.K. perspective
PRS

The Performing Rights Society was actually the MCPS-PRS Alliance up until 2009 when it became the umbrella brand PRS for Music of which the MCPS is still a part. The PRS exists in order to collect and pay royalties to its members which are made up of artists, songwriters, and publishers. It is important to understand that this applies to the composition/material itself: *the song*.

If material you have written or cowritten is recorded onto any format and distributed to the public, performed, or played in public, broadcast, or made publicly available online, then it is the PRS that will collect the royalties legally owed to you. For example, if someone wanted to cover one of your songs on their album, perform one of your songs at a gig, play your song at a club (DJ), or distribute your songs online then they, or the venue, would have to pay the PRS royalties in order to legally do this.

The PRS takes an administration fee from the monies it collects and the rest is passed on to the songwriter or publisher, whoever controls the copyright on the material. If you are a songwriter signed to a publishing deal then the publisher controls the copyright to your songs and will therefore take its own cut before giving you (the songwriter) the remaining percentage.

(Continued)

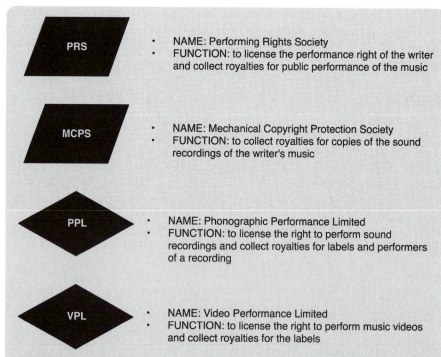

- NAME: Performing Rights Society
- FUNCTION: to license the performance right of the writer and collect royalties for public performance of the music

- NAME: Mechanical Copyright Protection Society
- FUNCTION: to collect royalties for copies of the sound recordings of the writer's music

- NAME: Phonographic Performance Limited
- FUNCTION: to license the right to perform sound recordings and collect royalties for labels and performers of a recording

- NAME: Video Performance Limited
- FUNCTION: to license the right to perform music videos and collect royalties for the labels

In the United Kingdom, there are four main royalty collection agencies licensing different aspects of the material's performance.

MCPS

The Mechanical Copyright Protection Society is actually part of the PRS for Music brand and is the organization that is concerned with the licensing of the song for reproduction on a recorded product.

If someone wants to use the *recording* you have made of a song on a compilation album, they would need to license this from the original label. (If you have also written the song then money would also be collected through the PRS.) The MCPS takes an administration fee from the monies they collect and the rest will be given to the owner of the copyright for the song. This is usually the publisher.

PPL

Phonographic Performance Ltd. (PPL) is the organization (in the U.K.) that collects money due to labels, artists, and producers, for the public performance of recorded music such as radio play, playback of music in public places, bars, restaurants, shops, and so on. It distributes the money to record companies and to the performers on the recordings.

Up until the late 1980s a producer would not have been recognized for her contribution to the artists' performances on the records she produced. However, members of the board of the MPG (Robin Millar and APRS' Peter Filleul) sought to gain recognition for the contribution producers make to the artists' performance on a recording. Subsequently, producers are now able to receive royalties via the PPL and are classed as "nonfeatured artists," therefore receiving payments when records they have produced are played on the radio or in public. (In order for this to occur, the producer must be registered with PPL and complete a database form with the information regarding the songs they have produced; this is where the producer registers himself as a nonfeatured artist.)

VPL

Video Performance Ltd was established in 1984 and was created to carry out the same role for music video that PPL was already carrying out for recorded music. Therefore VPL is the organization that collects money due for the public performance of music videos, such as on MTV and YouTube.

Record labels tend to own or control the copyright of music videos and therefore normally receive any royalties owed. However, contracted artists often receive a percentage share of VPL income from their record company under the terms of their record contract. Record producers may also receive a share of VPL income from music videos of songs they have produced, but again this will depend on the terms of their contract with the record label. (A producer would also receive PPL income from a music video as the recording they produced is used in the video; this would be based on the producer being registered as a nonfeatured artist with PPL for the recording in question.)

The American perspective

In the U.S., there are more than one main Performance Rights Organizations (PROs) for composers.

ASCAP

The American Society of Composers, Authors, and Publishers, commonly referred to as ASCAP, collects royalties on behalf of musical copyrights just in the same way as PRS does in the U.K.

BMI

Broadcast Music Inc., like ASCAP, collects royalties for their composers and publishers.

SESAC

Society of European Stage Authors and Composers, no longer just collecting for Europeans, is based in Nashville, U.S.A. and unlike most PROs is a for-profit company.

International collection agencies

For a list of international collection agencies please see Appendix F-1, The Tape Store at the end of this book,

THE TERMS OF A CONTRACT: PRODUCER (VS. WRITER VS. PERFORMER)

If you're hired as a producer, it is fairly clear what your role is expected to be, isn't it? Many might assume a rather puritan view, such as someone who guides the creative process, acts as a soundboard, provides musical suggestions, and coaxes great performances.

What happens later down the line when you find yourself cowriting much of the material with the artist in pre-production and yet you're not considered as a cowriter?

Perhaps you've not written the track, but you may have helped arrange the track in a new and more successful way and perhaps taken control of when instruments are played. Is this a cowriting credit or just an acknowledgment of an arrangement? Have you created the drums, the synth sounds, and sequenced your client's composition? If so, is this also worthy of a cowriting credit? These examples are probably not likely to gain a songwriting credit, but this will come down to a gray dividing line, one you will need to be firm and up-front about in negotiation. Be clear when you'll expect a credit and a share of the songwriting royalties as soon as you enter into any of the actions above. Take advice from your lawyer as to how this might be best achieved.

Many a disagreement has taken place in and out of courts over issues such as these. Be clear about this from the start or as soon as possible as issues arise and ensure your terms are reflected in the contract. Take notes during the negotiation and be clear about how to circumnavigate things that people may have innocently forgot or deviously omitted from the agreement.

Should you be awarded a stake of the writers' royalties, the writer's publisher might wish to negotiate deals such as synchronization and any licensing on your behalf. This saves them having to come back to you for permissions each time a request comes through.

RECORD LABEL TYPES

Majors

Major labels have been the dominant part of the music industry for many years. Well, in truth they still are, but there are fewer of them. In the power play that takes place in corporate business, consolidation and economies of scale often lead to bigger profits and seem a sensible pursuit. Most large corporations work in similar ways and the major record labels are no different.

There has been a tremendous amount of consolidation over the past decades to bring the whole industry to the "big four" that we currently know at the time of writing (Sony Music Entertainment, Universal Music Group, EMI Group, and Warner Music Group). For example, back in 2004 there were the big five labels but then BMG was merged with Sony to make Sony BMG in 2005 (now called Sony Music Entertainment).

This merger became a topic of conversation in the European Commission and while Sony and BMG were permitted to merge, it was not without considerable conditions. Conditions that included BMG selling off its publishing wing to its rival Universal, reducing overall market share.

These big four labels remain and it would appear that future consolidation, while perhaps sought by many a label's board, may be resisted in many territories because of market dominance and monopoly restrictions.

Consolidation, to some, might give the impression that there might be less business going on, or indeed fewer opportunity for acts. While this might appear the case presently, this is not simply because of the consolidation itself, but because of the new economic marketplace we find ourselves in.

Each of the four big labels has taken on a whole host of sublabels in mergers over the years and therefore have many subbrands, such as Sony's Epic label, for example.

The majors still dominate the majority of the industry. Many cynically believe these labels have become nothing other than large marketing machines more interested in taking on television talent show winners from the *X Factor* than developing real talent. There is some sensible business strategy in this. If the record-buying public in their droves votes for an act, becomes involved in an act, they'll buy the act's music. So perhaps for innovation and profits, this is a real business coup?

Submajors

These labels are distinct brands that the major label owns and trades under as if it was a going concern. Many brands have simply come and gone, being subsumed into the larger conglomerate label. However, there are labels such as Geffen which is now part of Interscope, whose parent label is the Universal Music Group. These brands remain and are often useful tools for the larger group to keep faith within the particular genres they operate in.

Independents (Indie)

Independent labels, or indies for short, have become valued places for bands over the years. These smaller, often more nimble, affairs have sprouted up from the ground promoting their acts well to provide a powerful alternative to the major labels.

There are some large independent labels out there doing very well in terms of the charts and success. Consider the well-known label Beggars Banquet which has had some current successful acts in the British pop charts in 2010. It has been so successful that it almost resembles a major label's conglomerate parent, insofar as it has become The Beggars Group and has brought together household music labels 4AD, Matador, Rough Trade, and XL Recordings all into one powerful collection.

An independent can be characterized by the fact that it owns its own business and is not owned or controlled by a major label[1].

Any label that is created by an individual for their own purposes could be considered an independent label too, but the following types require individual mentions.

Production companies

Apart from the standard independent labels as described above, others will also set up their own labels to meet their own needs and releases. One popular way forward at the moment is for production companies who develop artists and so on to create their own label to provide a much-needed conduit. To some degree, these have become valued

---

[1]www.musicindie.com/219.asp?sub=Join%20AIM accessed 13/08/2010//

(Continued)

content providers for the larger labels as they take on, sign, and develop acts. They resemble an external A&R agency in some ways. These acts, once the music and artist are ready, will become extremely interesting to the larger labels. In some ways, the risk is taken by the production company, and not the major.

Producers

Producers, in their own right, of course, fit into this category and are also taking the time to invest in their own artist development and exploring the many income streams that are possible from collaborating more closely with someone they might sign. This becomes an attractive option to both producer and artist as there is a need for them to work closely and with dedication to create a successful product. The DIY/360° type deals surrounding this are covered a little later.

Artists

Naturally artists often create their own labels whether that is just to get the wheels working for their own music, or whether it is for the promotion of artists they like. Madonna's Maverick label was one such business which successfully signed a young Alanis Morissette and launched her to the mainstream with *Jagged Little Pill*. This is a more common option for the future of the industry and many people, with the universal shop front the Internet provides, consider this the way forward. This way they can manage their own affairs.

Other: Live Nation and Starbucks

In recent years some interesting labels have popped into the marketplace, and gone. One such label has been that of Live Nation, which has "signed" acts such as Madonna to its roster. Their agreements are not for recordings per se, so they're not just a record label, but a more inclusive all-in style deal where all Madonna's income is accounted for.

Another strange entrant to the market, which has since disappeared, is the coffee chain Starbucks which launched its label in 2007. Soon after its launch it signed Sir Paul McCartney. Again there was some market sense in this proposition as "…you can reach 44 million customers per week through Starbucks stores," reported partner in the venture, Glen Barros, the president of Concord Music Group.[2]

The two examples above show something interesting. The industry is starting to look at new business models and ways of working. It is debatable whether both new labels have received the success they had hoped for. However, what is to be learnt here is that new business models may work, succeed, and break the mold. The industry will innovate and it will be interesting to see what transpires in the coming years.

Distributors

Distributors are the facilitators of the industry. They logistically ship the physical records through their networks to the shops or arrange the supply of music to the digital download suppliers. Many majors use their own distribution arm or have considerable contracts with an independent distributor. Independent labels can either make approaches to the independent distributor or make a deal with a major to distribute the records for them.

---

[2]http://news.bbc.co.uk/1/hi/entertainment/6445013.stm, accessed 18/07/2010

TYPES OF FORMAL CONTRACT

There are two types of formal contract that are of particular relevance to the producer: the *producer agreement* and the *production agreement*. The former could be described as the more traditional contract while the latter is a more recent, yet increasingly common, way of doing business.

In the next few sections we look at these agreements in their basic detail in order that an aspiring producer may know what to initially expect.

Producer agreements

Producer agreements are made between the record label and the producer for the production of an artists' record. Essentially the producer is contracted or employed by the label to produce and deliver a set number of tracks from the artist. These can be referred to as *master recordings* or *masters*. (A master usually refers to an individual recording on an album.)

The specifics of the agreement can vary and therefore the financial implications need to be realized and considered. As previously stated, the most important thing is to finalize an agreement before commencing any work on the record. These agreements and contracts provide an essential means of protection should the record label decide to move the goalposts halfway through the project or even start using another producer. Should this become an issue, evidence of the written agreement between the parties will be needed.

The costs deal

The *costs* deal (or *non all-in-deal* as it is sometimes called) works on the basis that the producer will be paid a production fee for the project but the label will book the studio and musicians and pay the costs. The average production fee on a major label depending on the producer involved could be somewhere between £2000 and £10,000 per track/master (at the higher end of the U.K. industry) or £300 to £1000 per track at the lower end. Veteran or top producers may command an even higher fee depending how negotiations go with the label.

It should be remembered that the label is footing the studio and musicians' fees and therefore this is not going to come out of the producer's pot. In other words, the fee is theirs to keep. Again, depending on the gravitas of the producer concerned and the outcome of negotiations with the label, this fee may or may not be recoupable out of future royalties. In other words, if the fee *is* recoupable then the producer will not start to receive any royalty payments until their fee is recouped. If the fee is *not* recoupable then the producer will start to receive royalties from the sale of the very first record. We should perhaps spare a thought for the artists here, as they have to wait until the total recording budget is paid back to the label (recouped) before they receive any money from the record sales. (An issue that a new artist may not fully realize initially!)

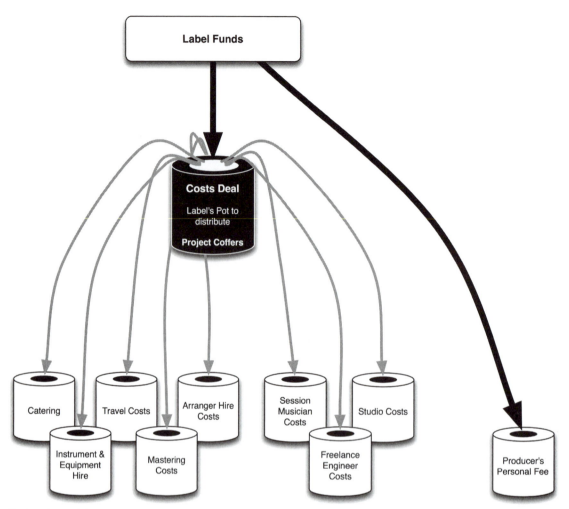

The key thrust of the costs deal is that you as the producer will receive a fee for the work you're doing, while the label will foot the bill (from the fixed budget) for the project costs.

The all-in deal

The all-in deal, as the name suggests, works on the basis that the producer will receive an overall fee or budget to produce the project; however, all recording costs such as studio costs, musician and engineer fees, and so on will be expected to be paid for by the producer and therefore come out of the same budget or pot. In this scenario the producer really needs to know how to handle a budget and costings for the various expenses that will be incurred during the recording process.

A budget will have to be submitted to the label to show the projected costings in order for an overall amount to be agreed on. A prudent producer will realize at

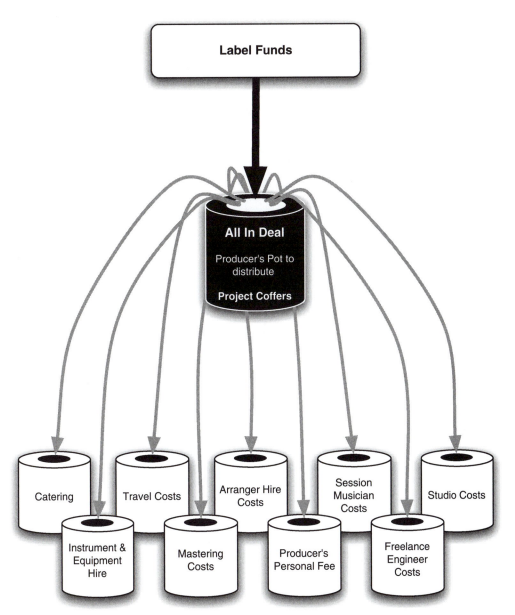

The all-in deal allows you as a producer to receive the money required to manage the production project. From this pot of money will come all the fees for studios, musicians, etc. as well as your own production fee.

this stage that some clever and careful financial management will enable them to save more money from their original budget, therefore leaving more to take home over and above their personal fee for producing the recording. A producer who owns his or her own studio facilities might opt for the all-in deal, as this may prove to be far more cost-effective. The same could be said for producers who have a good rapport and relationship with studio owners, as special rates can be agreed which will come in under the initial projected cost, which for budget purposes was based on the *normal* rates.

In terms of recoupable amounts the all-in deal would normally require that the personal fee or share taken by the producer from the total recording budget is the only sum that is required to be recouped from the future royalties of the producer. Any money that is left over in the recording budget outside of this personal fee is not subject to recoupment.

As with the costs deal, the recording budget and costs for the all-in deal have to be recouped in their entirety from the artists' share of royalties before they will receive any income from sales.

Production agreements

In response to the changes taking place in the music industry, more producers are now taking on the role of A&R talent scouts, finding new artists and working with them to develop their material and sound. Some will also take on a managerial role, organizing showcase events in order to place the artists in front of industry insiders and label A&R and making contacts for the artist or band. In some cases the producer may work with an artist on the songs, rewriting sections, or organize a band or session musicians for the recording of the project, and possibly even spend their own time and money recording demo or master-quality recordings. Enter the *production agreement*.

Production agreements are made between an artist and a production company (and can often be set up by the producer). This acts as a form of recording contract between the artist and production company or producer to make a record. If the project is completed successfully the artist and producer will agree to sell or license the recordings to a third-party label for release.

Setting up a production company and creating production agreements is the main way a producer can protect himself and have a greater degree of security than a simple verbal agreement. The production agreement may also help protect the producer's right to future income from her work.

The record labels benefit from these agreements in that the production company does most of the hard graft and develops the artist, leaving only the marketing of the album for the label to deal with.

The producer has a part to play in this. You could take on an act for a label on the understanding that you'll get a budget to record and work with the band, but you'll also get a point or two on the album. That's all very well if you're going to have a bestseller on your hands. Today you cannot always take the risk that album sales will make the income you'd like and therefore will need to spread it across other income streams.

THE INDUSTRY IS CHANGING

We were fortunate to spend some time in 2010 at a charity event with guest Sir George Martin. Andrew Marr was acting as interviewer for Martin's interview and relayed a question delivered from the audience, "What advice would you give someone making it in the music business now?" The honored producer simply responded "Don't!" to a hall of laughter. Sir George qualified his statement by saying "…that's unfair as music is more alive now than it ever has been. There are a lot of young people out there doing some really great stuff… try and create something that somebody else hasn't done… look for a different way of doing it."

He is not alone in his views in this current climate, as the industry goes through a period of organizational change. As we've established, traditional record labels are amalgamating and are in some form of decline as their market share becomes smaller. Independents are feeling the squeeze and small labels still struggle to put material out in a crowded marketplace.

Richard Mollet, director of communications at the British Phonographic Industry, illustrates how the mass industry has naturally declined in the U.K. over recent times: "This year (2009) only six albums sold over a million copies, whereas 10 years ago it was 13." Mass music purchase days are not over, but certainly lower than they used to be. Mollet adds that despite this, "trade revenues grew 1.4% in 2009, for the first time in six years. Part of that is because digital services got better organized and distributed… digital income is around 14% share of that."

Music to so many is considered free. Spotify, while as a company has had a tricky, often challenging, inception to the marketplace, has provided essentially free, relatively up-to-date, music to the masses. Why buy the album when all you have to put up with is a few adverts akin to listening to commercial radio?

It would appear that this made a major conversion in the psyche of the record-buying public. They no longer see a need to buy music when it's available free legally. And of course there are the illegal free copies (peer-to-peer file sharing, and so on). Surprisingly, iTunes does well from those people who want to listen to their music on their iPhones, iPods, and now iPads.

Richard Mollet ponders how new live streaming mobile music services might impact on current physical sales. "No-one knows what will happen with cloud music services and mobile devices. When this happens, perhaps in two years, who knows how this will impact on sales".

Technology will lead the way and will inform sales moving forward, but if music can be streamed on mobile devices, how might this affect the relatively stable marketplace of iTunes and the iPod?

So what's the deal for the producer?

What does this new financial landscape mean for producers and the industry as a whole? All the supporting machinery around the industry has taken a hit as the industry modernizes and consolidates. Many top-end studios that were once home to great acts and seminal, pivotal albums, have closed, while many struggle to stay afloat.

Why this has happened may be simply natural evolution, but not in the way that each generation seems to be a little bit taller than the last. Not in the way that one generation can prepare adequately for what is to come. This came rapidly.

The Internet and its websites and services such as Napster and other sites changed the whole concept of "free" music and the record-buying public bought into the idea, dispensing with the notion of the CD, and in turn perhaps even the duty of owning music. iTunes having been well-adopted, did fight to gain back legal market share, and continues to be a major player. However, so-called cloud services like Spotify now provide free listening.

The problem is that, for each CD an artist sells, they receive their royalty as contracted. Not a lot of money really, and the artist does have to shift a fair amount of units for it to count in this day and age. Then the producer has to get paid on royalties, if that's their deal. For each digital download version of their album they sell (of course, many listeners just buy the track they want and not the full album), there's a little less to earn per sale. Then of course the Spotify model is based on the amount of plays, and you need a colossal amount of plays to make it pay on an equal footing. For example, "Spotify users enjoyed more than 1m plays of Lady GaGa's song Poker Face—which earned Her GaGaness the sum of $167."[3]

The way forward

Given the changes in the industry, now the labels are wanting to see a finished complete product or at least a more rounded and developed one, thus offering them some kind of security that a risk (investment) might be worth taking.

This is why a new age of the production company is coming to the fore. If an artist has been developed, the material fine-tuned, and the rough edges knocked off, then there is far less for the label to have to work on (and therefore less investment of time and money). The production company may also have promoted the artist/band and developed a fan base, or following, creating the buzz that a label would want to see before taking a new artist or band on.

This new approach between labels and production companies is also starting to bypass the traditional role of A&R. The production company will now go straight to the label with their new artists and projects as they themselves are acting as A&R, essentially finding and developing the talent themselves.

As an aspiring or new producer, you might wonder where this leaves you. What is the best way forward for your work and career? Setting up your own or becoming involved in a production company may be one route that can really help

---

[3]www.guardian.co.uk/music/2010/apr/18/sam-leith-downloading-money-spotify accessed 14/07/2010

bring in experience and work. You may not earn a great deal of money at this stage, but if you are able to work alongside an artist as part of a production company and produce good results it may pay dividends for future work. A production company may have developed a reputation with labels for producing strong artists for further investment, in which case being a part of that team will bring positive exposure and experience. Of course, the natural progression from this involvement is to go it alone and begin to work independently.

The alternative to this is for you as a producer to attach yourself to a band in the early stages, almost acting as their additional member. This may be the long route to take and require much investment or your time (and/or money). However, if you are involved from the very beginning and become integral to the team, a label may be more likely to see you as part of the package and therefore involve you in any deals that may be struck.

A production company may bring in a "big name" cowriter or co-producer to work with the artist in order to develop their material, in the hope that the big name will add the necessary magic in order to produce some hit material. However, even a big name might not truly understand or get an artist and their material, which means that the collaboration will not necessarily bear the fruit that all concerned were hoping for. This is another argument for a producer being involved right from the beginning with the artist, where they help create the overall vision and fully click with the project. If as a producer you are able to do this, then you are potentially in a strong position when deals with labels are being discussed.

DIY movement

"You become a one stop shop which you have to do because of budgets these days. You can't afford loads of people... programmers and mix engineers, etc."

Tommy D

So how are the producers earning now? As we discussed earlier, many producers are preferring, or indeed looking for their side projects, to find and develop their own artists, and tie them into a decent production contract or, one stage further, working with them toward a 360° deal. Some might say these appear to be sordid and exploitative, but in fact can be written into producer deals with labels too these days as a matter of course. As we established earlier in the chapter, the producer is now looking to find their earnings from somewhere other than the recording as record sales are not where the money is currently to be had.

Tommy D, the British producer, believes strongly in the DIY upsurge. "If you find a great artist, don't bother selling it on to someone else. Structure it so that you can help them, so that they can get something great, which in turn will help you —it makes sense. [It's like] we've gone back to the '60s again with people like George Martin and Joe Meek... they found artists and chose or wrote their songs with them and then developed them. It's not like it used to be [say in the 1980s] anymore where you'd be sent some demos, meet the band down the pub and then spend three months recording the album with them, then see them play Glastonbury six months later. There just isn't the revenue anymore. Producers need to start thinking like Jay-Z and Puff Daddy... they're looking outside of the box."

Setting up your own label and production company alongside merchandising, brand, and the live shows appears to be the vital way to make money these days. As a producer, you should consider very wisely how you might involve yourself in all these income streams. As we began our interview with Tommy D we'd caught him looking at clothing wholesalers websites for "good shaped T-shirts for merchandising purposes," not in the studio as we'd have expected. Tommy suggests that producers might not be in front of a DAW most of the time these days, and might migrate that creativity to the social media websites to generate interest in their work and acts. Tommy adds though that "it [social media] will never take the place of the major as they have the marketing machine."

360° DEALS

360° deals have started to become more common for artists being signed these days. A 360° deal is one where all income streams of an artist are split up in varying amounts. This is often justified because physical record sales and digital downloading and streaming have reduced the income available from the recorded product. As a result, live income and other aspects of an artist's pot are being shared out.

In this type of deal, a percentage of the income take on the door on tour in addition to the merchandise could be sacrificed as a share of the distributed income. Similarly, any other income stream could be up for negotiation within the 360° deal.

360° deals (with a label)

While the 360° deal might seem incredibly unfair for the artist, it is an increasingly common and, labels would argue, essential way for them to recoup the investment in the artist and to bring future acts to prevalence through marketing and hype.

Much has been written about how unfair the 360° deal has become as the label strips yet more income away from the artist's dwindling pot.

360° deals (with a producer)

Producers are, as we discuss in DIY Movement later in this chapter, becoming increasingly more innovative and developing artists under their own steam. In order to achieve this and for them to recoup any potential income, they have embraced the 360° deal. The producer's choice to use the 360° deal is not one of greed or manipulation, but simply one to ensure they gain a return on their development.

Producers who are signing their own acts to their personal labels may be placing not only their time on the line, but additionally investing considerable money in ensuring the artist is developed. The 360° deal is discussed as an option to recoup this investment, so the wheel can turn for future acts.

Thinking about the 360° deal, producers that develop their songwriting skills, purely in focus of what will be a hit, partly because of their knowledge of the production industry, will place themselves in a better financial position. A share of the PRS income, MCPS and any synchronization those collaborated tracks might bring will keep the wolves from the door and keep his studio open. It's all then cyclical: if the producer keeps earning, then he can continue to develop new acts, which keeps the wheels of new artists' music flowing.

As we covered earlier, 360° deals might even be preferable for the artists, if they do not wish to have a large record contract around their necks for the future. Given that, it could be iffy that they will recoup any fees from record sales. A glum picture, perhaps? It has certainly got a lot of people thinking. What's the new industry model? Is it cloud services or is it live performance?

Now is the time to think cleverly about the deals you make, the work you commit to and how you can make the career as a producer pay. If you're likely to work with an artist who is going to be a guaranteed chart success, a normal points deal might work just fine. On the other hand, it is clear that if you stumble across talent and you think you can make it work, there may be definite benefits to keeping the work in-house—go for the DIY deal.

Musicians and producers will continue to have the thirst and drive for success and no doubt will sign a deal they should not have. It is important that musicians and music professionals do what they do for the love of it, but it is equally important that they are paid fairly for it. With the slices of the income pie being eroded from all sides, producers and artists should invest time and energy in getting the deals right. And again, we advise you to consult a lawyer.

See Appendix F-1, The Tape Store for links and resources about the legalities of the music industry.

CHAPTER C-2
Pre-Production

"The work for me really needs to go into pre-production. If the songs and the artist are ready, and we have a clear idea of what we need to do in the studio (including time scale), then the process of recording becomes less daunting and more relaxed, resulting (I hope) in better performances. Studios are stressful because they cost so much; the important thing in a way is for the artist never to feel that they are losing or wasting time in there!

"Once you're in and recording with my bands, things are almost out of the producer's hands, save just guiding the process, making certain decisions to keep things on track and within time and budget and keeping out of the way of the creative stuff. After that, ideas about mixing and mastering bottom out of many pre-production decisions and the recording sessions. I'm quite hands off."

Mat Martin, Hobopop Productions.

INTRODUCTION: WHAT IS PRE-PRODUCTION?

Pre-production, if you identify with the name, simply describes what should happen before a production starts, whether that is before shooting a film, or preparing for a West End or Broadway play. Following on from this comes production, which in music terms is considered to be the recording and mixing sessions clubbed together, despite often being discrete processes in their own right. Once the mix is achieved, it is moved onto postproduction, which in music terms this time means editing and mastering.

As a term, *pre-production* also can be an often misunderstood part of the process of making music and is often certainly underutilized. It's misunderstood because many of its facets, like music production itself, are not always easily described. For example, pre-production can represent the preparation for a recording session, including thorough rehearsals that could include some pretty advanced recording that may or may not appear on the final mix. Alternatively, for electronic artists pre-production and production can, to some extent, merge

What is Music Production. DOI: 10.1016/B978-0-240-81126-0.00007-X into each other as we'll discuss later. Producer and engineer Mick Glossop (Van Morrison, Frank Zappa) notes that "every band and project is different, but the principles [of pre-production] can be the same."

Arguably the process is defined from the moment a project is devised and comes into existence. Danny Cope, songwriting expert at Leeds College of Music, often refers in his work to the inter-relationship that exists between song composition and music production. Cope discusses a blend of the studio and the song, which has become a modus operandi for so many when beginning to produce records. Many artists record and produce as they go along in session, and for reasons we'll later explore, the divisions between pre-production and recording do not always remain clear-cut. In the example we explore, we are considering pre-production as a separate entity which can lead to incredible benefits, certainly in band recordings.

Decisions made at pre-production, or eventualities, may have immediately led to certain pre-production elements being applied that may need to carry on through to the final record, whether it be a specific sound of a unique synthesizer plug-in, loop, or the heavier use of rhythmical delay than normal on the vocals or guitar on which the whole track is now sonically hinged. In some cases it is the pre-production stage that will inform the performer or production team whether a composition, or an idea within it, will or will not work and hint at the developments that still need consideration.

Throughout this chapter we refer to the *traditional production process*, which signifies a model and nothing more of how we sometimes perceived the stages to be: clean cut and definitive. Once upon a time, these stages were devised partly through necessity and were partly defined by the composition process itself.

Compositions in the pop genre were perhaps typically based on a piano or guitar and subsequently etched into memory or were translated on to manuscript paper to be realized at a later date. The composition would then have been rehearsed with a band, or the manuscript passed on to hired musicians at the start of the recording session. By the simple virtue of this arrangement, the divisions between the stages were much clearer. The recording session was expensive, because of the sheer cost of the equipment housed, or even developed and commissioned by the studio in question (such as EMI in early days); and the range of personnel required for the session from hired musicians, engineers, and tape operators through to arrangers and the producer.

Recording sessions in the early days were often restricted to three hours as set out by the unions and it was assumed that the ideal performance could be captured within this time. Given this, the importance attributed to the preparation and rehearsals that make up what we can refer to as pre-production was paramount. An example of this was one of the Beatles' first sessions on September 4, 1962 between 7:00 and 10:00 pm in which 15 or more takes were recorded of "Love Me Do" and a number of takes of "How Do You Do It." Within these three

The Traditional Production Process Overview

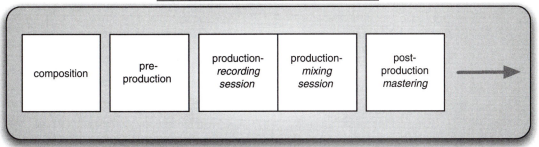

A Potential Modern Production Process Overview

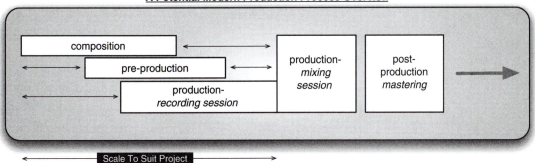

The process of production. The traditional model of the production process shows defined stages with no blurring. The latter model demonstrates the fluidity of the early processes in production. In this latter example, pre-production can blur backwards into the composition stage, and equally draws back some of the traditional roles from the recording stage.

hours mono mixes were made of each of the pieces (Lewisohn, M., 1989). This length of session could have been a demonstration session for management, or could quite easily be the recording session for release.

THE DEMO

The demonstration session, or the creation of a "demo," has been an important part of the production process over the years and has offered a tangible calling card for getting gigs as well as wooing prospective producers and engineers to work with. The demo should therefore be part of many a pre-production process as it also gives a recording for the band to listen to and reflect on in order to redevelop their songs, more of which we'll discuss later.

However, the demo is very much still alive and kicking for those bands not yet established or newly formed, as these will be the business cards for the musicians. The recording of demos is now so much easier to achieve using affordable, professional standard equipment. This availability of equipment has spawned a whole generation of home recordists who can provide that demo service.

The Production Process - including Demo Recording
This production process is familiar to many bands where demo recordings are needed to open doors and get gigs. This plays an important part in the band and song development.

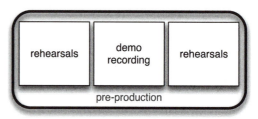

Many producers are said to consider the traditional demo dead for many applications. For example, established artists may feel demos are perhaps no longer necessary as they're working in their professional standard home studios. To some the audio may be considered part of the early sessions. It is presumed these original ideas may contribute to the final mix.

Smaller studios and suites are providing a good backbone for so much professionally released music with less reliance on the larger format recording studios. Modern producers and engineers are working within this new framework.

How these completed demos are used is variable. Most material is now data compressed to MP3 and posted on sites such as MySpace, which has proved very successful in sharing artists' music to the world. The demo itself has arguably expanded from a physical entity to that of a multimedia space including video links and audio. This is all part of the media toolkit that the presigned artist can consider before engaging with a label.

On a more practical and developmental level, the demo is, and can still be, a valuable tool to reflect on material as it is developing. There are downsides, in that when something is committed to "tape," it solidifies. Consider early demo recordings of songs you've tracked. Have they remained the same? Have newer versions ever lived up to the demo? Were the early versions not tracked so well, hence were unusable, but subsequent professional recordings have not made the grade and are missing some vital unbeknown quality you cannot quite put your finger on?

This is common. Many demos can solidify your thoughts and expectations on how a track should be shutting opportunity off for song development and this is where the producer will be able to see though this for the artist and steer the song in a different direction, which might provide it a new, or certainly, different lease on life. However, it is common also for these tracks to have something that is unique and special. Whether this is because the music was recorded when the excitement of the music was newly born, and was tracked quickly without too much thought to whether it would be released, is debatable. Suffice it to say that so often these demo recordings can provide elements of vocals, guitars and so on that have a unique, unrepeatable quality which we see transferred to the new version of the track. This is very often a great challenge, as the tracking is captured with less care and accuracy as it would be in the studio.

IS PRE-PRODUCTION NECESSARY?

Up and until 30 years ago or so, in the 1970s, the attitude toward making records had developed and changed to such an extent that an artist might have had the privileged opportunity to write, rehearse, record, and mix all in the same expensive studio. The financial advance from the label and potential sales for many artists were adequate to cover this, in addition to all the fabled lifestyle excesses we're used to hearing about. Attention to details such as planning, rehearsing, and arrangements for session musicians seems a far cry from this hedonistic example. Of course most sessions in this period were not all sex, drugs, and rock 'n' roll, we hasten to add!

Today, the studio availability has almost come full circle. The financial pressures are not so much that the equipment and quantity of personnel are the expensive commodity they once were, or that session availability is limited, but one of sheer business demand. For example, financial advances are far less than they once were, if they exist at all in some extreme instances. For this reason, managers and their artists are beginning to think much more frugally and business-like about how records can be made. As a result pre-production has renewed importance and value in that the production process can and should be carried out to maximum effect.

Tony Platt, engineer, producer and Music Producer's Guild director, says "[pre-production] has become more important for a lot of artists because it saves money on studio time. I still hear stories however about younger producers going to the studio unprepared and wasting studio time".

The lavish and expensive surroundings of a booked studio are usually ditched for a band member's home or a booked rehearsal facility, where concise planning and preparations can take place to maximize the outcome of the more expensive studio time.

As the times are changing, pre-production is one of the most fluid, hidden stages to the outside, nonmusical world. It is a process that can and should fit any artist and situation based on the desired vision of the product. Dependent on the type of song, album, and artist, pre-production can take many different forms that we'll look at in the next section. Pre-production is the bridge between the composition of the music and the recording session, borrowing skills and tasks usually reserved in other stages traditionally in the process.

FORMS OF PRE-PRODUCTION

Originally pre-production took on the form of planning and preparations on the part of the producer and assistants. This would involve the booking of the studio, engineers, musicians, perhaps even the catering, in addition to anything else that might be required. Attention to detail during rehearsals would be expected, due to the recording sessions often being short, with little or no time for experimentation. Any development of musical arrangement or structure would ideally need to take place in advance, in rehearsal. Perhaps the music might have been ready to record straight away.

Today, pre-production has evolved to a more valuable process. Producers take differing views on the importance of pre-production, some preferring to record whole hard drives full of takes from the rehearsal studio, to those who simply might arrive at one rehearsal for a quick listen to the band's material before beginning work in the studio. Tony Platt adds "Sometimes [I record rehearsals] but only very roughly. Spending time making top-quality recordings of rehearsals is a bit pointless. Although sometimes to capture that moment!"

Traditionally speaking the pre-production phase is often considered a vitally important part of the process. This is where so much new songwriting and song development can take place in the comfortable surroundings of the rehearsal studio, coming in at a fraction of the cost of a decent sized studio with an engineer. These are beneficial factors to consider in the current climate where budgets are more restricted. More on this later as we discuss planning a session.

More importantly, as Mick Glossop says, "pre-production is a very social process, as is the business as a whole. This pre-production time is where I get to know the musicians, the politics between them, and the way they work."

For those artists who do not rely on this traditional structure, pre-production opens up a great deal of opportunity. Many artists complete much of their work themselves at this stage of the process, leaving very little to be achieved in the so-called recording sessions. The stages of composition, pre-production, and recording blur together and adapt to the project at hand.

Andy Barlow, of the U.K. band Lamb, reveals that he creates music in such a way. For his music "the pre-production stage is not really separate from the creative task of song-writing, more it's the oil to the engine of creativity... [now] the only real time there is a big difference is if I am working in Ableton Live. I find this great for writing ideas and getting grooves down. Once this writing stage is finished I will then go ahead and [transfer it and] work on it further... in Pro Tools."

Despite Andy's command of the production process, he still prefers to involve specialists at certain stages. Andy's productions often include lavish sonic arrangements that require careful balancing between the occasional orchestral arrangement, real drums, acoustic instruments, and electronic samples and synthesizers. These elements all need careful mixing and, as a result, Andy prefers to pass on this responsibility to a mix engineer. "I choose to use a mix engineer for a few reasons: I don't like hearing the song hundreds of times [when mixing] as it clouds my vision of it. I get attached to certain sounds because of how long they took me to make, rather than whether they serve the song. I like SSL and Neve Desks (which I don't own). [However] I still like to be very hands-on with mixing, and make comments until I am completely happy."

As can be seen in Andy Barlow's example, the creative writing process of the song through to the point at which the material is mixed becomes one big pre-production session incorporating the writing of the material, and the development and intricate programming and production.

AN INTRODUCTION TO CREATIVE PRE-PRODUCTION
An introduction to song development

Developing a song to its full potential can take time and require specific ingredients. To compose a song and walk directly into the studio can extend the length of time it takes for the song to develop to its full potential, needless to say the cost to the project. In extreme cases it can simply reduce the overall quality of the end product. It depends on whether the song is reliant on the recording process and the studio or not. If studio equipment was required to aid the composition, the development and production of a song begins early in the writing process.

In our traditional rock band model, the writing process will perhaps begin by a member of the band at home or on the tour bus and be developed further when brought to the band in rehearsal. Let's dwell on the rehearsal a little.

It is in the rehearsal studio where much of this development and joint ownership of the material can be assumed. Band members are introduced to the new material and will develop new ideas including riffs, syncopations, harmonies, and so on, and even new sections. This is where a song can begin to breathe as part of the band ownership and not the sole composer. Perhaps this is the reason why there are so many dubious court cases about potential lost earnings from band member's contributions to the songwriting process.

Understandably, sole composers within a band often can be somewhat closed off and resistant to ideas that come from others. To exclude ideas from members of the band and outside can potentially close off exciting areas for the song to expand that should be embraced, if not simply given an opportunity to be heard. This will depend on the creative structure and internal politics of the band and how they develop their material best. The finer details of band interpersonal dynamics and their management are touched on in across this book and in Chapter C-4, The Desired Outcomes.

At some point in the rehearsal, it is worth bringing together the band and producer. The producer will be able to give a new perspective to the material and offer new ideas to develop the material yet further. Mick Glossop notes the important "arrangement changes, and choices of sounds" that producers can provide at this stage.

Similarly this can often be considered a threat to the composer's ideals and should not be seen as such necessarily. A producer's role is to ensure that the vision of the artist is successfully transferred to "tape" (sorry, hard disk, old sayings die hard) at whatever cost, within reason. As such it is important to embrace and welcome the producer's input at this stage; whether you ultimately agree with the ideas is up for discussion and rejection as need be.

Some projects will not warrant the cost of a producer. If costs do allow for one, perhaps there will not be enough funds to involve the producer in the rehearsal stage. Such input will naturally take place within the recording studio. However, for some unbiased commentary on the material, some artists call upon the ears of fellow musicians or trusted friends. This can give the band a new perspective or praise for the material as it stands, or some harsh, but true, criticism where necessary.

In addition, it is advisable to capture the rehearsals on anything from a Dictaphone through to a fully fledged Pro Tools rig, to take away the songs and listen in the cold. Mat Martin adds, "It's also imperative for me to know that the artist knows their songs inside out (I only really have experience recording singer-songwriters, and limited at that; I guess you'd approach a prog-experimental improvisation record differently). I do that by meeting with them months before going into the studio, demoing the songs, then pulling them apart and rebuilding them, tightening chords, lyrics, and structures so that they are more solid than before. I generally find that most songwriters think they know their work, but can't answer some really fundamental questions on it when pushed. Pre-production is about solving that for me."

WRITING TOGETHER

Often artists will choose to write and develop material with their producer. This has spawned many great albums and appears to be a fantastic way of driving music production in new areas of development. Marrying the producer, with the benchmarks of modern music making, with the new, fresh, and undeterred artist, can bring great new output.

Current indications are that more and more producers will become induced into the songwriting of the acts they work with. The main reason is financial, as we've mentioned earlier, insomuch as being a songwriter offers another income stream, often independent of the producer's fee, or points. In the world of digital downloads, and a fast-paced music industry, it is understandable that earnings from recorded music continue to dwindle in light of more widespread legal and illegal music downloads. Producers, such as Tommy D have chosen to embrace a modern way of working which he relates to "a return to the '60s" where it was common for the producer to be directly involved in the development of the artist and, to some extent, their income streams. One such example would be George Martin and the Beatles.

Many fantastic collaborations exist between the artist and producer, who knows the musical scene and fashion and can draw out the best from the artist and their composition. British producer Tommy D notes that sometimes songs do not complete easily. Sometimes there may be one idea which is very strong and can seem like a hit, but it may be lacking a strong chorus or middle eight. In this instance, it is worth shelving the part and waiting for inspiration to come together at a later date that will make the song whole. "It's the producer's job to come forth and say it's not good enough," he says. "The way to sell it is 'look, you've got a great idea here, but it's only half an idea. Why don't you complete it?'."

THE REINVIGORATED IMPORTANCE OF THE REHEARSAL ROOM AND HOME STUDIO

Now it's time to consider the way in which the songs will be performed and recorded. This is where we enter into arguably the proper phase of pre-production activity.

Again, this is the time for precise planning and rehearsal. Many bands still believe the recording studio is the point at which all manner of things can be fixed and made perfect: fix it in the mix, as the expression goes. This possibility of perfection to some extent is true with the onset of advanced tools such as Anteres Autotune, Melodyne, and countless others to tame even the most wayward vocalist. Add to this common tools such as Sound Replacer, BFD, etc. for making every kit sound as good as it possibly could do without moving a mic. Wrap this up within the awesome editing capabilities of all digital audio workstations (DAW) today, the performance can be shaped and crafted in whatever way you might wish. Taking this ability to "correct" a step further, the manufacturers of Melodyne and Celemony Software have created *direct note access*! (We thought it was an April fools joke at first, but this feature claims to be able to separate each note in an audio performance for editing in both time and pitch.) This allows incredible advantages to the producer and engineer, but this cannot always make a bad performance good, and in the cases where it can improve a performance, it will not necessarily make the track as a whole better. Nothing can beat a solid performance as we'll talk more about in Chapter C-4, The Desired Outcomes.

As we've already mentioned, the rehearsal should still remain an important opportunity to improve the structure and arrangement of any song. The emotional architecture within the arrangement, using dynamics, and harmonic content should be scrutinized and adapted to suit the performing musicians and their foibles. The performances of those musicians playing on the track bringing their own ingredients or killer bass lines to the melting pot should be assessed and developed. As such, the rehearsal, with a producer, becomes an important point at which the song can breathe and take shape.

The rehearsal stage should also offer the band the opportunity to tighten up the performance of the material to ensure that the recording session is smooth, ordered, and efficient. Some bands might wish to play the album material live as a whole set, just to try out the material and to play it solidly in front of an eagerly listening audience.

REHEARSAL RECORDINGS

Recording the rehearsals can allow the band to hear how they actually sound. It is so often the case that the musicians themselves get a different perception of their performance near their amplifiers on stage. So often the band can lose the perspective necessary when constructing and producing a song.

Producers can vary in their approaches if involved in the rehearsals. Some will bring down a small studio set-up to capture the musical events. Often, material in the rehearsal studio can have a quality that, for some reason or another, simply cannot be recreated later in the studio. Whether it is an angst-ridden vocal take or a particular sound to the guitars augmented by the room, the amp lying around the rehearsal room and player's grubby strings at that time. All this serendipity should be captured. It used to be so often lost to a substandard recording format such as a cassette based 4- or 8-track machine, the quality of which was normally less than adequate for a final release. As the cost of equipment has reduced, obtaining high-quality digital recordings of rehearsals has become readily available.

Producers often develop this recorded material and edit it to demonstrate how a different arrangement or parts could improve the music. Development of the material can take many different forms: whether that be messing about with the structure while sitting alone with an acoustic guitar, to moving parts in a digital audio workstation. This can then be replayed to the band to rule out indecisions that individuals are discussing. The proof is in the listening.

Either way, the song will have the opportunity to be structured, to be developed, eventually to breathe.

PARAMETERS FOR CHANGE

The alteration and recomposition of elements such as melody, harmony, and so on are the parameters by which the song development will take its form. The methods and order by which these developments happen cannot be prescribed, and the producer, or other objective listeners, will possibly base their opinions on a gut feeling. They may hear a different sound or arrangement and suggest changes accordingly. Song development is hinged on a number of criteria from a personal list of things that people look for. Typically the key overarching aspects a producer, or reflective artist, would wish to alter on an objective listen back to their material might be:

Arrangement
Instrumentation
Tempo/time signature/groove
Performance quality

These four overarching parameters can offer a great deal of scope when altered. Listening objectivity (discussed in Chapter A-2) to music can present a list of things to repair or could spur new ideas as required.

Rehearsing with the band in a loud rehearsal room can be an exhilarating experience and can make material seem solid and workable. However, an objective listen at a lower volume while using a framework can produce a list of elements that will require attention.

Depending on this list, experiments and alterations to the arrangement, instrumentation (acoustic or electric), groove, and so on can be tried and might be recorded to listen to later. Having a decent recording system within rehearsal allows you to jam around ideas freely and on listening later, elements can be edited together for experimentation.

RECOGNIZING PHASES OF PRE-PRODUCTION

Pre-production has its own internal phases. Phases can be recognized and worked with to monitor progress in a production. Understanding these phases can provide unique points from which to consider the more mundane aspects of preparing for a session.

Musical progress

Phase 1 begins with the songwriting itself, which we assume is a hive of successful creation. The next stage within Phase 1 is the point at which the individual's material is presented to the rest of the band. The progress may feel as though it is hampered to the composing member, as the band are learning parts and arrangement and beginning to enter into the next stage. This is the initial development of the individual's music by the band as a whole (band-based song development). Parts will be forged, riffs and arrangements created, and songs developed as a whole.

During phase 2 many artists might begin recording and use the studio time to develop their music along a similar path. Phase 2 assumes that the band objectively listen to

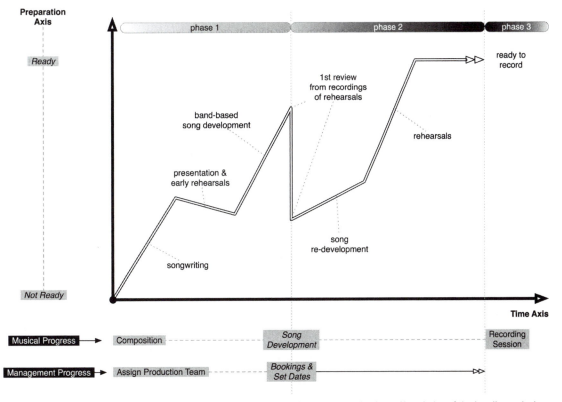

The perceived progress of a band's material can be mapped as in this fictional example above. Knowledge of the band's musical progress can perhaps also act as a trigger for management activities within pre-production, such as booking musicians, studios, mastering, etc.

their material, either through some kind of recording of the advanced rehearsals or through traditional demo recording sessions. It is at this stage that comments and constructive criticism will be revealed and might provide an action list. This list, it is assumed, will cause the artist(s) to step back and perceive that their progress has been hampered, or set back slightly. Either way, this is an important reality check for the artists as they return to the drawing board a little and potentially carve out a new direction, or alteration to bring the material into line, whether that be a change in tempo, key, or indeed time signature. This will not necessarily take a considerable amount of time, but will ensure that the material is simply ready when the red light goes on. Ensuring that each player knows their parts and can perform in "second nature" mode, not necessarily having to think about every note, will result in a faster, more productive session.

The band will develop and practice the music until it reaches a plateau of regulated performance quality. In other words, the musicians will repeatedly deliver a consistent and engaging performance of the music they're planning to record either in the rehearsal studio or live performance. It is at this point that the band is ready to record and enter into Phase 3.

Management progress

If a band's progress can be tracked, then it is possible to predict when the material is in its second phase. It is at this point that the music becomes galvanized into something new using the information provided on the list, or reverts back to the better original format in Phase 1. At the start of Phase 2 the material is brimming with new ideas and may be forged into its final state. It is during this time that studios, session musicians, and other bookings could be made with some form of certainty as the format begins to settle down.

This certainty provides an opportunity for the artist, producer, and production team the opportunity to plan Phase 3, the recording session, properly. It is at this point that musicians will be confident of the elements they will wish to include in the songs, and therefore the equipment, personnel and space needed to produce the music in question.

THE ABILITY TO LISTEN OBJECTIVELY

To make an accurate assessment of music, it is important to develop certain skills in listening. To be able to do so in an objective manner is imperative before developing songs or choosing to throw them out.

We all listen to music for enjoyment, but it is extremely difficult to do so in an unbiased and detached fashion. As Danny Cope outlined (Hepworth-Sawyer, Ed., 2009), a song should have influence and naturally draw the listener in. Thus to measure this influence, the art of listening objectively should be developed, as it is difficult to remain focused on how the music will appear to the listener when taking part in its development over time.

As we discussed in Chapter A-2, Analyzing It, it is useful to have the ability to switch between detached (passive) and attracted (active) listening. This is a skill that is useful to be in a position to shape music accordingly. This should be something we should be able to do for a particular audience whether we like the material or not. Objective listening can take time to develop. Audio professionals such as producers and engineers spend a professional lifetime listening with an objective ear.

In a similar fashion to that of a classical composer who can imagine or emulate the sound of the orchestra in their head, many audio professionals can also hear music altered and edited, engineered and polished before any equipment has been operated. This ability can be developed and is something that first begins with concerted analytical listening and then can be switched off into detachment as required. This detachment means the listener analyses sound on occasion as opposed to music.

Working in music and audio brings along an allied objective ability. Strange, but useful, is the ability to listen to the same song, verse, or guitar part over and over again without tiring of it, whether we like the piece or not. Being able to detach yourself from this may seem a little like masochism to the casual listener, but is imperative to work as we do, while attempting to remain objective.

Later a model for planning is outlined for a production, and listening skills such as these enable us to identify whether or not the material is on the same road as we had planned. The framework discussed in the next chapter may not be detailed enough to guide the listener to immediate objective qualities to make or break a track. Nevertheless, an improved listening ability should ensure that problems in a composition or its arrangement could be identified and rectified.

WHITTLING DOWN THE MATERIAL

Choosing which tracks are to be recorded can happen at various points. At the end of Phase 1 to the start of Phase 2, the band may listen on reflection to some of their material that might be a joy to perform, but as a standalone song will not engage with the fan base. More recordings can be made of the songs throughout Phase 2 to assess the worth of pursuing the material. In most cases these songs will move forward in the kitty to the session as if/when they are recorded they will be of great use as b-sides.

So many bands work on many more compositions than make the final tracklisting on a CD. In many cases it is not uncommon to rehearse and record many more songs for a CD than the typical 12–15 tracks. Working with a producer, or taking on the duty as a band, the material will begin to get sieved and perhaps at the same time, the singles will begin to present themselves, perhaps alongside a track order, as the material gains a persona of its own.

This whittling down of material is an iterative process and takes time. Thus adequate planning should be given to returning to the drawing board after the first phase of the pre-production process. When this point is will be dependent on the band and the composition team within the project.

The V.I.S.I.O.N

A structured starting point should be employed. For this purpose, V.I.S.I.O.N. has been produced. This can prompt the team into considering a number of aspects about the project to ultimately agree on the vision for the product. This trigger list is the halfway house between the creation of the project and the project management to come.

Going straight into a studio can be an excellent way of getting started and developing music. It is fair to say that most would all love the opportunity, but as we've outlined, this is not always possible depending on the size of the project. As such, some direction is useful to ensure that an end goal is achieved. Succumbing to your business mind will ensure you plan accordingly.

Any such plan must follow a path that has been agreed and mutually understood by all concerned, otherwise how would we know we had got there or which road to travel on? To plan toward an end product, a vision needs to be considered, or at least the anticipated outcome needs to be visualized.

To assist this, we've developed a simple checklist called the V.I.S.I.O.N. that addresses many of the aspects that could be conceptualized before beginning a project. Securing the vision of a project will empower everyone to talk from the same hymn sheet. It is important that this vision is easily disseminated to others and understood with as little misinterpretation as possible. Below we look at each area individually.

VISUALIZE

Visualizing the end product is not necessarily just a dream of how the material may sound, but should also be a target position in its relative marketplace. An understanding of the current music scene and how this will slot in will be of use when considering the material and what direction it should take. Consider that of any genre that is constantly evolving such as hip hop or dance music. These change very frequently and visualizing the music's position within that marketplace will give pointers to where it should or could fit, as well as the suitable inspiration to differ and innovate.

INNOVATE

With each record that is released, innovation is often at the core and musical genres develop almost as fast as computer processing power growth as stated by Moore's Law. While innovation might not be the sole aim of the game for the artist, it will, by its sheer virtue, be something new. Whether this is a radical departure for the band and thus alienating its audience a little, or whether it plays it too safe to keep hold of the fan base, it will still be new material. The million-dollar question is how a band can generate enough interest album after album. New music needs to inspire and draw listeners in as described by Danny Cope's triangle of influence (Hepworth-Sawyer, Ed., 2009). Creating this interest

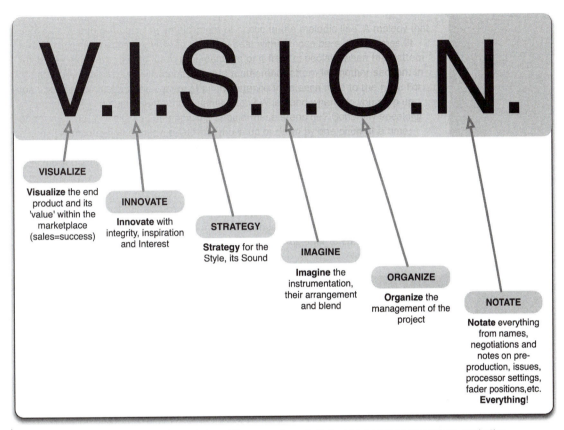

The V.I.S.I.O.N. is an acronym we use to remind us of the many elements that might need to be planned pre-production.

can be drawn from inspiration, but can be delivered when working with a specific producer. Thus, the producer can have a profound influence on the music.

Being in touch and having an ear to the ground musically will influence a different and fresh perspective to the material. An awareness of the genre and scene will also provide a much valued success gauge. Thus, knowing if this tune was to do well in the charts and clubs or not, or whether more work would be needed to get it to the same standard or exceed it, would be valuable. As such, it is not simply how it sounds but its value within its genre or marketplace. This is connected with the style and sound of the work, which we cover a little later.

That said, it is important to consider the origin of the material and treat it with integrity and not follow a personal route. The fellow band members or producer should take into account the wishes of the composer and try to bring the material to life without altering it away too far from its origins without agreement. Sensitive musicians will present some of the hardest challenges within the studio and integrity to the music and the cause is an important pursuit for all.

STRATEGY

Based on the value judgments and assessment of the genre and scene that have been made, it is important to produce a strategy for the style and sound of the recording. This strategy will be an outline of the kind of sound you're wishing to achieve and will point you in the right direction in the organization and planning of your project.

Style in this example is meant not only to be about the sound, but could be the way in which the band looks and acts. The charts are usually full of what is known as "manufactured" artists that have been styled already. This is an important facet and clear selling point, and in some cases it will begin new trends in society for dress sense. This style may not simply be about the clothes that are worn, but the way in which the artists carry themselves. For the average producer working for a label-found artist, this will not be too much of a concern. However, in the current times, more and more producers are starting to act as A&R, promotion, and artist development in addition to their traditional role.

IMAGINE

Imagining the end product is one thing, but noting the instrumentation that goes up to make the vision is crucial in the planning. Much of the instrumentation may be dictated by the songwriting itself and be somewhat dependent on the parts, such as a riff-based song. To picture the blend of the instrumentation together can lead to new sounds and ideas, which can lead to new production flavors.

At this point it is also worth considering arrangements of music pieces and their position in the track listing as songs. During production this will invariably change, but it will assist the planning and the list of things the band need to achieve before recording and beyond. For example, a concept album will require to work in a certain order, such as *The Wall* by Pink Floyd.

ORGANIZATION

Pre-production will involve a healthy amount of this, as we'll discuss later in more detail in Chapter C-3, Project Management. It is during this stage of the production process that the vision is determined or crafted, rehearsals are about to start or are in swing, and the planning needs to happen in earnest based on the vision for the music. To begin to consider the production project as a whole and how it will come to market is important at this point. Choosing the production team, the studio, and so on, is all part of the planning process.

NOTATE

Many bands and indeed studio professionals fail to notate as much as they should. Bands may forget older songs that have fallen by the wayside. Without records, these compositions could be lost. The same works for the logging of settings, agreements, conversations, ideas, and notes. Disciplined logging is rather useful in planning large projects. It's your business, after all. Many of us fail to keep up with note-taking because we do not have a system. We discuss this more in Chapter B-3, Being a Business. Note-taking can come into its own when you need to recall agreements made between people within the studio or elsewhere. You may need to refer back to this from time to time.

Engineers and producers should also take notes about all important effects, processor and compressor settings for a big mix session they're trying to recall. Notation such as this is far easier these days with recall on pretty much all modern equipment, but still, there will be the odd coveted Urei 1176 or Pultec EQ that will need to be marked off on a template somewhere. Alternatively use a digital camera or your iPhone, unless of course you use something like Teaboy (*www.teaboyaudio.com*) for notating all your outboard settings.

Notating the event through video or photographs is also something that can be used for the band's or production team's website, or career materials for the future. If they never make it into the public domain, they will be something to view with interest in the years afterwards.

V.I.S.I.O.N. PROMPTS

By using the prompts provided by the V.I.S.I.O.N., answers to prominent questions such as these below should be forthcoming. Alternatively, these questions should at least prompt the team to consider these areas further until a solution is reached.

Visualize

- Is there an agreed visualization, or feel for the end product?
- If not, how can this be reached?
- Is there a concept behind the album; story or just a collection of songs?
- How will this product fit within the genre now; in six month's time?

Innovate

- Are there any benefits to adding some new innovations to this record?
- What could these innovations be?
- Will they alienate current fans or will they act as an influence for new listeners?

Strategy

- What strategy can be put in place to create the visualized sound that is wished for?
- What reference points can be drawn to achieve this end product; sentimental music with comedic lyrics about Mickey Mouse?

Imagine...

- ...the instrumentation and consider this as part of the overall visualization. Mainly a drum'n'bass beat with operatic vocals? Perhaps heavy metal guitars, bass and vocals and a tambourine instead of a kit?
- ...the arrangement, or again style of arrangement.
- ...the sequencing (the order of the material on the album).

Organization

- Can you begin to consider the planning of the project?
- Who are the production team to be?
- What rooms and studios can you use or do you want to use?
- Will you need external musicians and equipment?
- What planning can you make at this stage about the initial release?
- Is there a specific date you'd want to release this, such as Christmas? FA Cup? World Cup? Olympics? April Fools Day?
- How do you want the product to be presented based on the answers to the concept and visualization above?
- Artwork ideas?
- Packaging?
- Management of the digital downloads?

Notate

- Can you confidently remember everything? If not, create a notation scheme you trust so that planning and thoughts can be captured.
- Prepare notes on all aspects of the production that pertain to you.
- Can you prepare notes about the material for the production team?
- What else can you notate and provide for the rest of the project team about the music and the project as a whole?

CHAPTER C-3
Project Management

Making an album can be an organic process given enough time, money and space to be creative. In these days of ever-increased financial pressure, the time, money and space to be creative is dwindling fast. Producers need to be ever-more efficient, effective and economical.

Producers are often thought of as creative flamboyant people, with the authority that can guide music to become a hit. Actually they're also the people that make the project run and the sessions succeed practically. As we mentioned in Section A, there are many types of producers and their involvement will differ dependent on the project, personnel, and music accordingly.

So here comes the caveat: some producers will wing it, taking it as it comes. You just have to do that with dwindling budgets and creative ways to achieve the same things on less money. However, to some extent or another, the project will need to be project managed by the producer or one of his or her employees. That could be on the back of an envelope or, because of the complexity of the project, on some sophisticated software. What we provide here is an idealized account of how you *might* go about it. There are no hard and fast rules, just suggestions.

In corporate circles, project management has become big business in recent years with some very respected professional training courses emerging, such as PRINCE2 and the Association of Project Management's range. These courses are detailed and exhaustive and, while a lot of the principles within the programs might be of benefit, music productions require a very different skill set and sensitivities, but also share some common values.

If you're already a producer with plenty of projects under your belt, this chapter will definitely be one to skip over. However, if you're still learning, as many of us always are, then this chapter will either have some gems as to how to improve your project management, or provide inspiration to research further systems that might help you in the future. The things you learn here regarding project management can be used for your current music project or anything else for that matter.

What is Music Production. DOI: 10.1016/B978-0-240-81126-0.00008-1

PLANNING FOR SUCCESS

So often the word *planning* seems to grate against most people's view of what record production is. Planning and project management are mandatory for businesses and many walks of life. It is a serious and expected skill. One would not take on a business venture with a large budget and various personnel involved without considering the legalities, financial position, and meticulous planning. So why would a music production venture be vastly different, providing enough leeway is reserved for the creative process? Project planning can take many forms and is explored in this section with reference to music production specifically.

Planning the project from start to finish might seem a very dry and restrictive thing to do, but it can be a necessity to some extent. It is common to read about sessions with the Beatles in Abbey Road in the early days lasting a few hours, but the sessions from later albums seemed to last for weeks on end. This is clearly an exaggeration, but most of us will not get similar opportunities in today's commercial studio and because of this there needs to be a determined end point. A delivery date with a tight budget.

The world has changed a little and things move at a considerably faster pace nowadays. Combine this with portable and high quality audio equipment expanding what is possible away from the control room, and there simply isn't the need to stay and develop so much material in the studio anymore. This could be taken in two ways. It might suggest you take less studio time so that you can work in a smaller room (perhaps your own), only outing to the larger rooms for drum tracking. However, the other way to look at it is that you as a producer need to make the process as efficient as possible. One way to achieve that is through serious and meticulous planning.

As hinted, so much is based on the finances of a project. The traditional, larger record labels have less money to throw about at each of their artists. Worse still, the artist might see themselves bearing the costs more and more as new models of distributing music become apparent and record deals are slowly adapting. As such, planning takes on a new, more poignant importance to ensure that the project comes in on budget, let alone the negotiations.

All projects have driving parameters for completion; most often these are tight deadlines and even tighter budgets. These two parameters have an overriding effect on how the project should and can succeed. These could be seen as restrictions, but can also act as serendipitous opportunities in the sense that the design of the plan can lead to creative new thinking that could give your project an edge over the competition.

Some projects will take next to no planning whatsoever as the artist may have their own studio facility and like to play all the instruments. In this situation, it is simply a matter of planning when to get in their studio together to crack on. There will always need to be a plan as you, the producer, are likely to have prior commitments for the next album you may be working on. In the same vein, the artist may have touring or promotional commitments too that need to be factored in.

However, many projects will simply take a little more time to plan and organize based on the complexity of the music in question. Projects that need to include session musicians (see Resources Management, below) need to be factored in, as well as any arrangers and orchestras you may wish working on the tracks.

In real life, of course, each seasoned producer will have a way of working that will be fairly flexible, might include a personal assistant or a system that simply works. However, for those early-stage producers among you there are many ways in which you can look at project management. You could map out the production as though you were doing it day by day using software that produces time-line diagrams called Gantt charts (which we'll mention a little later). There are also other tools available to help you, even using an online calendar could prove to be beneficial at mapping the project out over a period of weeks. However, before we delve into this, we ought to look much more at what needs planning and suggest some very business-like ways to deal with them.

WHAT SHOULD BE PLANNED?

Have you ever sat down with all good intention to plan something, whether that be a book like this, or what equipment you need for a session, or the clothes you need to take on the next residential recording stint? Do you find yourself sitting there thinking "Why do I need to do this?" or "surely this is not necessary"? Well most of us do, and this is the reason we don't do it.

This is the strange dichotomy with planning anything—it seems so unnecessary. However, really how often do you think "I wish I'd remembered X or Y" or "How on earth did I forget my Mac Book Pro power supply, the nearest Apple Store is 200 miles away." Well we're simplifying it a little. As the project gets more complicated, with many people involved, the planning becomes less irrelevant and much more a true necessity.

So what follows is in part rare in the type of session you might frequently experience. However, we've chosen this as a description of some of the typical considerations you might have. We've also been idealistic with respects to the amount of time and supposed budget you'd be allowed.

As you read this you need to consider how the industry is changing, and how new and innovative ways of tracking, mixing, and later releasing are changing the time-honored traditional way of doing things. We discuss this in Section E, The Future in a little detail.

Planning whether to use the traditional label, healthy budget and rock band example below, or a modern equivalent, the skills of project management are universal and can be employed to assist efficiency in any line of work.

Shared vision

So what should we be doing when we plan for a session? How do you know you've got the right content to fulfill the project from start to finish? A good place to start is to naturally ensure you have a shared vision with the artist. In Chapter C-2, Pre-Production we spoke about the vision in a little more detail and what this means to the success of a project.

Once you have a clear idea of what you want to achieve with the project ahead, you can visualize the end product. With this in mind, you may have a clear sense of each of the elements required and how and when they'll fit in the jigsaw puzzle. Also you'll be able to translate your rationale to your artist, and collaborators a little easier if you understand each other.

Forming a plan of action

You'll need to prepare the studio, the band, perhaps also the meals, the accommodation, the orchestra, oh and the arranger, the engineer, his assistant of choice, the mix engineer, and the mastering engineer. Have we forgotten anything?

The whole gamut of things you may need to consider, book, and pay for may be, at times, overwhelming. It is understandable why so many busy producers do indeed make use of fantastic administrators with honed project management skills. Of course, again this is assuming you have this level of project to deal with.

We've put together a list of considerations here for planning an album project. We've worked on the assumption that every aspect needs to be covered by you or you'll get the artists' manager to do some of the personal bits.

Take this scenario based on a good budget and plenty of time. It's based on how some albums were made a while back. We've taken this decision to show you how to manage and prepare for a pure project, without too many modern restrictions. Thinking about the project management on this level will free you to consider the order and rationale of each item being where it is and the music production at heart. However, in the real world, the principles of the time management will be useful, but the album plan might be much more restrictive and require you to think efficiently and creatively about how to complete the work ahead.

A SCENARIO

You're taking a band to a residential studio in the countryside to record an album's worth of material. You've got a fairly relaxed deadline of over a month. You've got to complete the album mix in this period too.

Having been in pre-production with the band you now know the material well, the band's strengths and weaknesses. From this you've decided that you'll want to do an orchestral tracking session on one of the ballads back in London before the mix sessions begin. This will mean the four weeks' (approximately) worth of sessions will be interrupted. Planning all this may need some careful thought.

First of all you'll need to consider what you're recording and when, and if there is a reason—why! What to track when and where can be an interesting place to start. In rehearsals/pre-production is there a song that you think is more complete, or well-rounded than others? Is this a good place to start? In some ways you could use this track as a warm-up for the album. If you think there is something good and solid about this song, the performance from the band is tight and accomplished, and it has a good vibe about it, this could then set the tone for the sessions to come. Psychologically, having one rocking track recorded and that sounds great might be fantastic to get people in the right frame of mind.

There is naturally a converse side to this insofar as this might give false signals to the band that the rest of the session is going to be as easy as that. However, good communication should allow for all to understand the rationale. You'll measure this with each artist, because every project is different.

Making decisions about what to record when and where might appear restrictive and in some ways counter-intuitive to the creative process. In some ways it is, but if you already know that one of the hits requires strings, then you'll need to schedule the arranger, the orchestra, and the large studio to track them (in an ideal world). The deadline will not necessarily move because you want this on the album, so it's a matter of squeezing.

Therefore this knowledge might make you consider that you need to track this song a little earlier in the sessions, so you can pass a rough mix to the arranger to work from, giving her adequate time to produce a score. This is one simple example of how the order in which you record the album can take shape.

So you may have planned out enough time for all the songs you wish to record and, in some rough way, when. This might get scribbled on the back of an envelope, even a fresh piece of paper, or more eloquent techniques used such as a Gantt chart or other project management software.

Slotting in the session musicians so they arrive at the right points in that schedule will be important. It is fair to say you may not get the idea for that wacky flute player until halfway through the recording, at which point you'll get on the phone to your contacts. However, if you know what you need in advance and because the session player is rather sought-after, you'll need to fit in with their schedule too. So booking early can be essential. Thus having an idea of when you need people and keeping to your schedule is important.

PLANNING FOR SERENDIPITY

A word of warning as you plan should be pointed out here. Stress in any form of work can come from a deadline looming which cannot be moved, and you cannot see how on earth you're going to get all the stages completed in the time you have left. Sound familiar?

As familiar as it is, it can be handled in two very different ways. Whether we'd like to create a plan to avoid this or not, the deadline will loom and a mild panic might ensue.

(Continued)

Out of this can come some inspired decision-making and some serendipitous results, providing some memorable recordings, even hits! However, in reality this stress will not always guarantee serendipitous outcomes and sometimes the lack of time, with the stress it brings, may provide an unhealthy environment to produce music. It's a constant balancing act.

Contingency in any plan is essential, whether that be building a house (how many public building projects go over time and over budget?) or a new large IT project, and is one major aspect of project management. As you begin to plot out the album sessions in front of you, contingency should always be considered for many reasons.

Studios are creative places, where musicians (and producers) can think and be creative with the musical raw material at hand. Carving out new sonic sculptures and arrangements from the songs can take a little time and, however well-prepared in pre-production, new ideas will begin to appear, which could come from your comments, your artist, or engineer. You may wish to spend some time exploring these ideas and in doing so put off a tight schedule. One over-produced song will not compensate for the rest being an under-produced album.

Injecting some contingency time into your plans will allow you to experiment safely in the knowledge that the rest of the album can be recorded with a similar level of contingency. Some might say that this is a waste of time and resources, which it may be, but invariably you'll use every second of that extra time whether that be on the mixes or overdubs, and so on. We must note though at this point in time, that budgets are dwindling and as a result the pressures of time are up, unless you think creatively. Do always make sure that you've left enough time for file backups, file management, and any CDs/DVDs that the label might want burning off. We touch more on this in Section D-1 The Session.

Contingency is not always about the creative. There will be times when the equipment, through no fault of your own, or the studio for that matter, will fail and there'll be some incurred down time. This is to be expected and while the studio might be happy to extend your stay, it will not necessarily move your deadline with the label, and as is often the case, the studio might not have the available time in the studio you're using when you need it, throwing your majestic plans out.

How much contingency you allow for will come with experience. Too much will mean you pay for a pre-booked studio when you did not need to, which is not necessarily great for the budget, but is better than too little. On many projects a rule of thumb can emerge. For a business one of the authors was involved in, it became necessary to quickly work out how long a new job would take and while we might say a week, what we really meant was three times that (adding contingency and down time). The same, incidentally, was about right for cost (and the client in those days was happy with an invoice that was less than the quote).

Time contingency has an added advantage too. A less stressful studio can make the process far more enjoyable and memorable. Being up to the wire constantly might be the impetus required for some musicians to get the album done, while for others this could be an unwelcome distraction.

It is worth considering the life and experiences of the average professional band. These people are often on the road. They may not get to spend a great deal of time in the expensive home they've managed to buy from their first few successful albums. The demand for tours from their fans and managers (as this is where the big money can be for the artist these days), can be relentless. Many musicians might welcome the creative freedom of a relaxing, if protracted, rural recording session, assuming the label or band can afford it.

Getting contingency just right is an art, but always make sure you feel comfortable you have enough of it!

Session musicians are only one part of the jigsaw. Obviously you need to make the booking with the studio, the accommodation (if it's not part of the studio complex), catering (again if it's not contained within the studio), and travel (the artist's manager will take care of that you'd hope), among many other aspects.

During pre-production you may have spent some time working with your artists and the songs that have come forward. In an ideal world, as we discussed, the artist will have more than simply a number of songs needed for the album and will have provided much more to choose from. In that instance you'll need to make the right selection before hitting the studio. Indeed you may record more tracks than is needed on the album in the knowledge that you can decide later on, dependent on how the sessions go. Any extra tracks you record will, for the band, make extra b-sides and exclusive tracks, while for you, will increase revenue dependent on the deal you make.

Once you've decided on the tracks to be recorded it is whether you place one song in the recording schedule in front of anything else and if so why. There may be no rhyme or reason for you to choose one before another if they are fairly easy to track and can be taken in any order. However, for those tracks that need session musicians, or different studios, then you'll need to give them a fixed point in the schedule to arrange the external factors required. Once you've booked you should be ready for them.

That sounds reasonable enough, right? Well, next comes the real guessing game. How long does it take to track a hit single? You might be able to guess that accurately perhaps. However, you'll never really know how musicians react in the studio. A colleague once mentioned that "the studio is like a microscope" referring to the precision that people place on their performance. Knowing this and the syndrome it creates, musicians can overanalyze their playing to the distraction of the recording. How long will it take for the guitarist to nail his solo? Will the band get the backing track done in a reasonable three takes or three hundred? Will you need to compile (comp) 10 takes together to make one?

Many of these questions you may be able to answer because you've spent time in pre-production with them. You'll be able to accurately assess whether they can play, whether they're performers and whether the songs are any good, or will need some form of studio trickery. From this you may be able to make informed guesses of how much time each song may take to complete.

At this point you may wish to make a plan of action graphically, or just tot the amount of days required for each track and play around with the order based on when you think it would be best to track certain things. You want to start with a surefire track that the band enjoy playing live, and gets their crowd going. They already know it, they can play it and it rocks. It, should be a good place to start.

Moving tracks up and down the schedule will become clear as you start to plot what you might like to put on the track in terms of additional instrumentation. You might work something out on paper like the calculations below. Note that this schedule has been made in a linear fashion, tracking each song one by one. We call this serial recording.

Obviously on an album project like this, especially where the band are well-rehearsed, you might choose to economize by efficiently tracking all the bass and drums to all the tracks in one go (with scratch guitars and vocals for later overdubs). If the band can play, then this should take no longer than a day or two. This we have called *parallel recording* (not to be confused with parallel compression).

Working out the number of days in serial recording sessions is one thing and, as you can see by the table here, the producer has not provided much in the way of contingency should the schedule start to slip (see the sidebar Planning For Serendipity). Moving the tracks that need some sorting out earlier up the schedule might be sensible, especially "Seven Eleven" which will need an orchestral part.

Once you and the label have committed to something so grand (as it's very expensive), you'll need to ensure the session is firstly booked and the orchestra is free. Next you'll need to ensure the music you're asking the orchestra to play to is recorded soon to provide the arranger/orchestrator with as much time as possible. Therefore, this will need to come up your schedule. Remember, you'll need enough time to bring the orchestral recording back to the studio and feed it in enough time for your mixing sessions.

So you've got your songs listed and the amount of time you think each one will take to track. You realize that you've been a bit mean on time and realize that to do things properly while working serially, you need to add enough contingency. How much is a good question and it's best to start with an arbitrary figure such as a day per song, if you're lucky and you feel that is necessary. For some bands that you have worked with before or that you are confident can play the tracks live and well, this contingency will be pure luxury, so you'll need to make the appropriate judgment.

The way in which you approach the recording of each project will be different. Some bands will be excellent live performers and will be able to track the drums, bass, and perhaps some of the guitars alongside a scratch vocal in a few takes. This is a really efficient way of recording that we refer to as parallel recording. Serial recording is where each song is dealt with individually and more or less completed before moving on to the next. Of course you can choose to use a blend of the two methods should this be best for the project.

This is where it can get complicated. You can, of course, keep a lot of the information in your head, or swim along hoping all will be well, but you'll need to think logistics at some point. The logistics in question are who, when, and where. Which musicians do you need and at what time? In some cases, where the sessions take place at different studios because your band are on the road, the "where" will also need to be answered.

Song	Additional Players/Studios/ Notes	Estimated time to record?
"Song One"	N/A – they'll nail this in one or two takes	1 Day
"I'm Always Coming Second"	Extra flute player required	2 Days
"Why Do Buses Always Come in Threes"	Horn sections required	3 Days
"Four to the Floor"	Collaboration with DJ	3 Days
"Take Five"	N/A, but might take a little bit of time to track tightly.	2 days
"My Six Speed Car"	N/A	1 day
"Seven Eleven"	Ballad – would benefit from orchestral part – contact Brian Morrell for Arrangement	3 days in studio 1 day in Abbey Studio 2
"Eighth Grade"	N/A	2 days

Song	Additional Players/Studios/Notes	Estimated time to record?
"Nine Lives"	String Quartet – call Brian for arrangement. Also Band have not worked this one out completely yet	4 days
"Ten"	N/A	1 Day

This table shows how you might have to think about the tracks you're about to record. Despite the titles, these tracks are in no particular order. Note the additional aspects such as orchestral sessions and consider how these need to be managed in your schedule. Also note that studios will need booking. Has this producer given enough thought to contingency?

Do these sessions need equipment that the band own and haul from studio to studio? If so, who do you tell so that it is in the right place at the right time? Do you need to hire specialist equipment such as the Mellotron you want to add to the band's version of "Take Five"? Does the studio have the right bits of kit for this session or do you need to hire in your trusty analog 24-track machine?

How you think, organize and track all these aspects of the sessions will be up to you, but we believe you can manage it with your PIM and a lot of thought, or you could pass the work over to your PA. However, you could employ something that creates a Gantt chart, using project management software. Something like OmniPlan (*www.omnigroup.com*), for example, provides a comprehensive set of tools to manage a project such as an album project.

Below is an example of how you might arrange the planning ahead of the album. Note the three days of global contingency given over to allow for slippage within the project.

Where tools such as this become powerful in their own right is in managing the time of each member. This would allow you to note clashes, should the bassist have a family engagement they cannot avoid, such as her wedding, or the singer being interviewed by a national television program about the new album. You can place this in the schedule and the system would note that your best-laid plans to track vocals for that track need to be altered a little to compensate.

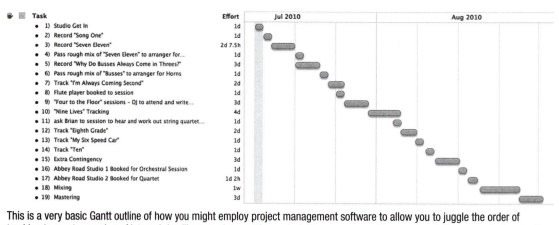

This is a very basic Gantt outline of how you might employ project management software to allow you to juggle the order of tracking to meet a number of internal deadlines, such as getting material to arrangers on time, or completing songs in time for the orchestral sessions you have booked.

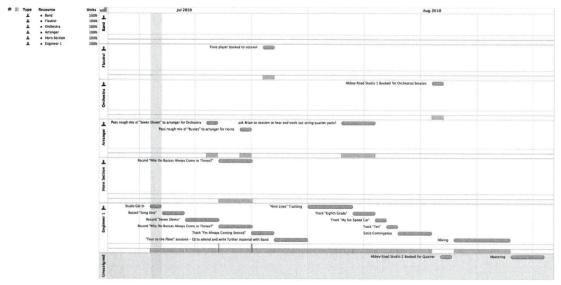

Project management software such as OmniPlan can offer you the ability to see when people will be required for parts of the sessions. The system can also flag up clearly where clashes might arise.

In the interests of keeping the diagram relatively clear, we have not littered it with the minutiae such as the bassist's wedding and any complex travel arrangements. Nevertheless, you can hopefully appreciate the benefits of planning in this way.

Taking time to concentrate the plan in some detail with travel, personnel availability, studio availability, and so on will ensure a smooth running set of logistics. However, as we mentioned earlier, it is not always necessary to go to all this trouble. You will have to gauge what you think is necessary for the type of artist, the money being put into the project, and what the outcomes are to be.

Resources Management & Financial Management

Some projects that are homegrown and emanate from a digital audio workstation inside a project studio may not have quite the same time constraints and can therefore be more flexible and resourceful than a large recording studio production. Having said this, timings, as discussed above, can still be important in terms of ensuring the material is written and arranged perhaps in time for a booked session with session musicians.

Resources management is a large aspect of project management. Calculating when session musicians, equipment, and so on are required at certain points is relatively simple enough. Although as we've established, estimating enough time for these requests will be a skill that is developed over time.

Ensuring the technical and personnel resources are available at each eventuality could be key to the success of the production. Working in a feature-rich studio with 100 U of outboard will probably ensure that the production team is adequately supported in terms of equipment. Personnel might cause different challenges.

Session players can be expensive and can work on strict time limits. Some musicians follow union guidelines and are often only booked for three-hour to four-hour slots at a certain cost. If you have asked for an orchestra or even a simple string section, then you will wish to have the track in the correct state for the start of the session. This helps your scheduling as it allows you to accurately decide how long you've got to track something. However, it is the cost of something that will be the deciding factor, especially in current financial times.

THE BUDGET—PROJECT COFFERS

Managing the budget is one of those regrettable administration things you'll have to do as it relates to the session, unless you have an artist manager who is keen and wants to get it done himself because he's funding it from an alternative source (relatively unusual).

How you have determined the budget will have resulted from a number of things as we discussed in Chapter C-1, What's The Deal? You're in a strange position insomuch as you may have negotiated a deal with the label, but the money you will be spending is not theirs to deal. That is because the money will be the artist's money recouped from the sales of their records. In effect you're spending someone else's money they have not yet earned. Thinking morally about it, it can be hard as you plan the best opportunities for the music to shine.

You'll have different income streams on the table as per the type of project and deal you may have with a label. Alternatively, you might have decided that you think an artist is worth investing your own time in, so therefore there is no budget other than what the artist can pay you or what you're prepared to put in for free, perhaps at your own studio. (See "Production Agreements" in Chapter C-1, What's the Deal)

That same old chestnut comes up again: contingency. Always be hard on the budget and how much things will cost, as invariably there will be costs you've not foreseen, or that will crop up. Not accounting for them in the budget, yet needing a special pedal, tape reel or whatever, will mean they will have to come out of your pocket (your profit).

Given the discussions in the previous section, quite a lot of expenditure can come out of what is an ever-dwindling pot these days. Finding a studio that is fantastic value for money, has great facilities, and will accommodate your band in the country might be a taller order than it once was, given that there is less choice in studios in some territories.

In the same way we plotted out the time it would take to track each song, you should also consider how much money each track would take and add it up. This is perhaps over-simplifying the project a little, but you'll be expected to consider the best way to spend that money, and also try where you can to make a profit in the case of an all-in deal.

FINANCIAL CONSIDERATIONS

This is where we press Apple+Q on Pro Tools, Logic, or Ableton to open up Excel or Numbers or another spreadsheet package. "Did they *really* say spreadsheet package in a music production book? Have they gone to the dark side?" Yes, we did say it, and for good reason.

Spreadsheets are simple, yet effective, in ensuring that your project is managed financially. What is spent where and when is important to keep hold of. Naturally there are a number of different solutions to the financial software package, including some dedicated packages (not necessarily for the audio industry though).

Just like any professional, or household budget for that matter, making a spread-sheet that lists every possible estimated cost will really focus the mind on the money available. Place the amount of money each item will cost you. Leaving a column for you to put actuals against later can be very useful when planning both this project and similar future projects.

On your spreadsheet include everything from the session flautist, the extra Otari MTR90 you want to hire in, and the Mellotron. Add also any costs to do with catering, transport, hard drives, other media (tape for your MTR90), and so on. Yes, you'll need to consider this, unless the deal states otherwise. Go and make the right deals with a local restaurant, or the in-studio catering to keep costs down.

MAKING THE FINANCES FIT

This should give you a figure which will not exceed the budget being provided for you to get on. However, invariably in the current economic climate it is quite possible that you may exceed this and have to make cutbacks appropriately.

What	Estimated Cost	Actual Cost
Studio Time		
Engineer (freelance or part of studio time cost)		
Assistants (freelance or part of studio time cost?)		
Session Musicians		
Additional Studio Time (tracking orchestras/overdubs etc)		
Arranger		
Media Costs (Hard Drives, Tapes, CDRs, DVDs, Wax Cylinder)		
Equipment and Instrument Hire		
Travel		
Catering/Food		
Accommodation		
Misc Costs		
Mix Engineer, Studio and Equipment		
Mastering		
Producer Fee		
TOTAL		

Roughly this table would need completing with the relevant figures for your session. Costs for your studio, musicians, and so on will vary considerably depending on where you are in the world.

Earlier in this chapter we provided a rather idealistic example of how you might want an album recording stint to go, working serially track-by-track. As we mentioned, this is not the most efficient way of doing things for all artists. You may choose that, as the rehearsal studio is much cheaper, you wish to ensure you spend as much time there ensuring they can play tight enough to employ a parallel recording style, getting all the rhythm sections down one after the other. This will make costs come down slightly as you hope things can be tracked just that little bit quicker.

Also consider recording in a well-appointed studio near where the band can live in their respective homes, saving heaps on accommodation costs. You'll probably need to put yourself up! All common sense really.

However, the other aspect to making the savings you need would be to spend a little of the money you've received on your own home-based studio. To do this properly with acoustics and so on, you'll naturally need a fair old chunk, but to place gear in a reasonable sounding room and crack on with some of the overdubs may not be such a bad move financially.

In the longer term, you might wish to invest in your own room, or rooms, as you can legitimately charge the label or client for studio time. Except this time, it's into your pocket, not the studio manager's. Also this might not be such a bad move as you can save a fair bit of time traveling, and therefore money. You will need to decide whether you want to bring that work home with you and all that entails. Realistically, this is the way things may go in the future, leaving larger rooms for the larger things, such as drums and orchestras.

Making the finances fit a pre-determined budget will often be a challenge, and is one that will require you consider whether to send the mix to one of the big boys, do it in-house yourself, or entrust your engineer to do it at home. Similarly you could choose to cut costs by not using a leading mastering engineer, and choosing a lesser-known one. It is not always advisable, as the label may insist that the "hits" are at least mixed and mastered by two of the small legions of top-end mixing and mastering engineers, which many labels believe make hits happen.

UPDATES

You've submitted the budget some time ago and you're now in the thick of the sessions, and what you thought would go well is taking a little bit longer than expected. That might be because the band play well live and feed off their audience and somehow do not gel in the studio, or simply because there was not the money in the budget for you to spend protracted time working with the band in the rehearsal studio.

If the schedule slips and so does the budget, you would be wise to meet with the label's A&R as early as possible or talk on the phone to discuss the issues. Explaining what is happening and what solutions you've considered will trigger thoughts in the label which could come back as a bad mark against you, but if you've been honest with the label all the way through, they may appreciate that you've said you need more money. The label may have to reconsider their position.

Sometimes you'll have to resort to fixing deficiencies in the mix, although this can usually take so much more time than simply performing it properly in the first place. Nothing replaces a top performance!

CHAPTER C-4
The Desired Outcome: Strategies for Success

INTRODUCTION

As we have discussed earlier in this section of the book, preparation can assist in making an album go smoothly in the studio. It is possible that it can also increase the cohesion between band and producer, and thus it is hoped that a better outcome can be realized from the whole project.

Much analysis can be made of friendship, human interaction, ego management, emotional undertones, and body language but the process of recording must simply happen and there are some mechanics that need to be considered.

In this chapter we discuss some potential methods and considerations to assist the flow and ensure that the important ingredients are given equal attention in the recording process. We also explore the need to balance the creative forces of music making with those of the financial and business obligations as laid down by the production. It is important to consider all the ingredients of the music and prepare for them accordingly.

BALANCING THE FORCES

Within this text we are at pains to remind the reader that music is a business just like any other and should therefore be taken as seriously. However, the mechanics are not the same and the achievement of a successful outcome does not come about from systematic and precise processes. Mainstream business does not allow for bad days where creativity is not forthcoming. The recording industry is reliant on the creation of a product, which is subject to considerable influence from many forces from the songwriter, band, singer, producer, engineer, A&R people, and managers, to name a few. There is a significant balancing act to perform to ensure everyone is happy and is prepared to compromise. This is where the skill of the producer comes in.

Balancing such forces within the creative process should be seen as the producer's greatest skill. Can a balance between the fluctuating creative process and the mainstream business model of planning be sought? Indeed, an individual set of rules and project management is drawn on for each project.

What is Music Production. DOI: 10.1016/B978-0-240-81126-0.00009-3

The timeline is not set in quite the same way, but nevertheless, the studio time and release date may be finite and within that window a record needs to be produced. A seasoned producer will know what works and how long things are likely to take as we mentioned in Chapter C-3, Project Management.

For example, a string session will be a short but exhausting affair simply due to the sheer cost of the personnel involved and the union structures that limit session times. However, a studio get-in and set-up might be a more relaxed affair. No session can necessarily be planned to the finite degree, but with adequate contingency, based on previous experience, deadlines can be met.

VISIONING THE PRODUCT

In many forms of project management, the visualization of the finished product is given an important mantle, as we discussed in Chapter C-2, Pre-Production. Can you visualize the final product? If so, can you work the steps back to plan for a successful outcome? It is proposed that through systematic and detailed planning, an outcome can be achieved. This planning is common in management and business. Institutions such as the Association of Project Management (APM) define what project management actually is and run courses in how to plan such things. However, is it right and correct to assume that music production can follow the same processes?

The methodology above might be appropriate to producing a physical product such as a table or chair, but in such a potential ephemeral product such as music, involving people, sensitive people, it is virtually impossible to plan that meticulously. Nevertheless, this does not mean that you should not try to lay down a vision of what the product will sound like. This vision will be something that is based on your watermark, unless of course you wish to reinvent this.

The vision may not account for taking steps back to plan for a successful outcome, but nevertheless it is important to have a picture of what you hope the album to sound like before embarking on the session. A great example of this can be found in Robert John "Mutt" Lange's production of the Def Leppard album *Hysteria* (described later in this chapter).

There have been many recordings where the artist has been so intertwined with the writing process that the vision has entirely been theirs, with little room for the traditional producer. One example of this would be Pink Floyd's *The Wall*. Roger Waters on the whole wrote the seminal double album, with some of the more popular tracks written with David Gilmour such as "Run Like Hell" and "Comfortably Numb." *The Wall* was entirely Waters' vision held around personal experiences. However, it can be argued that Bob Ezrin's production sound is very clearly heard on the album.

Roger Waters had always enjoyed telling stories of one form or another and this is why his subsequent album with Pink Floyd (*The Final Cut*) and his solo record, *The Pros And Cons of Hitchhiking*, both tell stories and should be consumed on a holistic level as one piece of music. One could argue that these pieces were more story-led. Whether or not Waters had a complete vision for the music on both albums from start to finish is unlikely, given the improvisational freedom afforded to Eric Clapton on *The Pros and Cons of Hitchhiking*.

One way of achieving an end result keeping to the vision would be something called plotting.

PLOTTING

Planning can give way to something that's referred to as "plotting" by some producers and engineers working in the studio. Plotting can take place with regards to either the song itself or the production. This is essentially where the production team will take one or two other songs and use selected elements or characteristics in order to plot out their own production. This is certainly not a new concept; in his 2009 book *PWL from the Factory Floor* (WB Publishing), Phil Harding describes how "song and production plotting" was a frequently used term in the 1980s within the PWL team and was a common way of working on new productions.

Phil writes:

> It meant that you could use another song to allow you to influence yourself to write a new song in the style of. Often two different songs might be used so as to lead the originator and listener away from its origins, i.e., you might use one song to plot (or copy) the groove (feel of the drums/bass and percussion) from and another song to plot the chords/arrangement and melody from.

Before we assume that this method of working is specific to that of commercial pop writers and that plotting only happens within the same musical camps, we ought to mention and consider other writers from slightly different areas of the musical spectrum. Barry Manilow, Eric Carmen, Andrew Lloyd Webber, Billy Joel, and more recently, Muse have all in one way or another allegedly taken inspiration from classical music composers. In some cases the ideas and influences are more obvious than others, such as the band Oasis; having a careful listen to some artist's work can lead to some interesting similarities!

Production plotting is similar in approach to song plotting but varies slightly in that it is the *sound* of a track and/or artist that may be plotted and referenced. At a very obvious level, consider the auto-tune vocal effect that seemingly dominated the dance and pop tracks in the early 2000s. This can be traced back to the track "Believe" (1998) by Cher, where the over-use of the Antares Auto-tune software resulted in an interesting vocoder-like vocal effect. It is rumored that the "telephoney" sound heard on Cher's vocals in this track came as a request from Cher herself, having heard a similar effect on a Roachford record. Again, production plotting!

Song and production plotting is still very much alive and kicking and although it may not be so openly talked about, it is a frequently used method for working on new material, especially within the commercial pop genres. We only have to listen to the pop charts to realize that the current trends and sounds are being replicated by more than one or two artists or bands. This idea and way of working does put a slightly different slant on the visioning of the end product. The question is not so much "How do we want this record to sound?" but more "Who do we want this record to sound like?"

PERCEIVED PRODUCTION

If you listen to songs from the same genre or style, there will always be many similarities for obvious reasons. However, we usually perceive that each song has been individually produced and therefore should present something original and unique, or certainly individual.

This is where *perceived production* as we call it comes in. Perceived production is not a million miles away from production plotting. We may perceive that a track has been produced (which obviously it has) but the production is simply following a stylistic template or pro-forma as it were—a pro-forma that has stood the test of time, more or less untampered with through the ages.

If we think of the singer-songwriter genre many aspects of the production are generic across a multitude of artist's tracks, especially the rock ballad style.

The beauty of perceived production is that the listener can somehow anticipate what the next part will do, what the next drum fill will be, and how the chorus will resolve. It is somehow like an old childhood favorite song covered by a new act of the day, but is actually a new, completely different composition.

We perceive what general instruments will be playing, when they will come in, the type of melodic or rhythmic part they might play based on the general norm for that genre. We therefore perceive that creative production has taken place when in actual fact it is possibly nearer to emulation. We perceive the production.

THE PRODUCER'S GOOD C.A.P. (CAPTURE, ARRANGEMENT, PERFORMANCE)

Music production can be crudely dissected into three aspects. We refer to this as the *producer's good C.A.P.* (Capture, Arrangement, Performance). We say "good" as, by rights, a good arrangement, with ultimately a good performance captured well, will offer the best basis for a solid recording. It is worth noting at this point that the original songwriting needs to be of a significant standard. Additionally many creative activities can take place to yet further the production of the music, but the following ingredients deserve the appropriate consideration.

The producer's attention to what we refer to as *good C.A.P.* should be extremely high. With good arrangements, excellent performance allied with great sounding and sympathetic capture, it is possible to produce and get a good sounding recording. Many productions have ultimately relied solely on these ingredients. It is a relatively recent phenomenon that the studio has become the wizard in the process. Yes, many things can be doctored in the studio, but without a good arrangement and individual performance, the music will not reach its true potential.

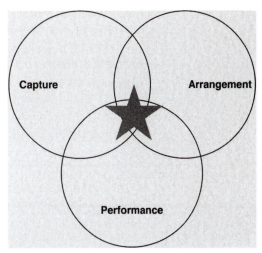

The *producer's C.A.P.* refers to a toolkit that could be adopted to ensure that the ingredients are balanced to create a musical delicacy.

Despite excellent recordings and performances, many producers and engineers "fix" and arrange the outcome of their recordings by editing the takes together to make the mix more consistent. This is a widespread practice today and, allied to other processes such as auto-tune, is the norm. However, it is worth noting that this kind of editing is exceptionally time-consuming and the better the take, the less editing required. This is good on two counts, one on your production finances, and the other that the creative process can flow faster for the band. Nevertheless, this should not detract from attempting to gather a excellent performance in the first place, hence the good C.A.P.

The importance of these individual pillars of the production process is arguably only second to the way in which they integrate with one another. The relationship between *capture*, *arrangement*, and *performance* is something that should not be underestimated as together they form the essence of the production. A strong or successful production is often one where these three elements work together in a balanced and sympathetic way. If one element can be judged as weak then the others will suffer. Conversely, if one of these areas has received more attention than the others then the same effect occurs.

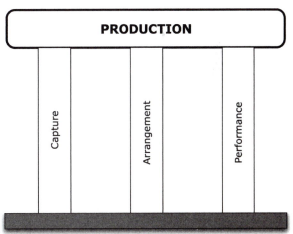

The three pillars of the production process. Together they form the infrastructure of the production.

The vision of the end product will of course have a huge influence on how these elements are approached and executed, and differing genres will certainly not require the same contribution from these three elements. However, before commencing a recording it would be wise for the producer to consider their own production C.A.P. methodologies and preferences, and consider how these will influence the recording process and whether they match the artists expectations. In the next sections we consider these three elements of production in the order in which they would generally occur.

Arrangement

Although *arrangement* appears as our second letter in the producer's C.A.P., it is most often the element that comes first in the production and recording process. The term *arrangement* can conjure up a whole realm of different meanings and interpretations.

For many, arrangement means a notated string or brass part, for example, a string or brass arrangement. Others might consider the choice of instrumentation used to be arrangement and further still some would consider arrangement to be the organization of a song in a structural sense, for example, verse, chorus, mid eight, chorus, outro. Whichever seems most applicable to you, it is important that as a producer you get to the bottom of what arrangement can be and not what you might think it is. As one of the three pillars of production, the arrangement is a major benefactor to the success of a song.

You may consider yourself to be a producer or an aspiring one but would you also add "arranger" to your résumé? Likewise if you consider yourself to be an arranger what would make you stop short of using the term *producer* when referring to your skills? Is there a difference between the two roles and if so what is it? For a moment let's consider what an arranger and producer actually do. You might see yourself in a slightly different light.

TWO SIDES OF THE COIN: PRODUCER AND ARRANGER ROLES—BY BRIAN MORRELL

Brian Morrell has written extensively for TV, radio, theatre, and the Internet. During his continuing professional career as a piano player, composer, and arranger, Brian has played a role in some of the most successful shows staged, including *The Phantom of the Opera, Chess* and *Aspects of Love*. He has worked as a session musician and arranger for some of the 20th century's top pop figures, including Elton John, Eurythmics, Sting, Natalie Cole, Barbra Streisand, Kate Bush, and Gloria Gaynor, to name only a few. Brian has lectured in music theory, composing, arranging, piano playing and history at many different colleges, mostly in London. He currently holds the position of Senior Lecturer in Music Production at Leeds College of Music, U.K.

The sound of a piece of music is something that transports to the listener immediately, usually before the emotive, harmonic, melodic, or lyrical impact of a song. T.S. Eliot said "true art communicates before it is understood". He wasn't talking about music production but the phrase applies and underpins the profound power of orchestration, arrangement, and production, its uniqueness and its comparable qualities.

The most important issue to grasp when discussing the similarities between producing and arranging is that both roles are informed by art and technology and in reality are separated only by history and interpretation. Producers, like arrangers and orchestrators before them, have always worked at the cutting edge of available technology, but always with one eye on the art. It was never about technology for technology's sake. Production is about harnessing creative skills, using suitable technology creatively to manage sound and music; so is arranging and orchestration. Both roles have always embraced technology, but only for what it can bring to music; and what it can add to the art.

Essentially arrangers and producers have always been the same person, divided only by subtly different skills. The result is the same; the process slightly different. The advancement of production has benefited traditional arrangers greatly in that they are freer to write more creatively in the knowledge that any subtleties and abstractions they employ can be dealt with sympathetically by a producer who understands orchestration to a degree. Indeed as an arranger in today's world one must possess knowledge and understanding of production in order to grasp the potential of a recorded arrangement. Although the concept of production sprang out of the shadow of the arrangement, production has informed the arranger in many ways. Arrangers cannot and should not live in ignorance of the possibilities production brings, any more than producers can resist the need to understand more about instrumental and textural subtleties.

Arrangers and producers share not just the basic skills needed to truly understand the management of sound and music, but also some of their industrial applications too. For example, there are those who specialize in studio arranging; who produce stunning arrangements and vivid orchestrations which only have to be played a handful of times, so can be adventurous and demanding. An example of this might be the vivid and abstract orchestrations of George Fenton, contained in much of the BBC News themes of the 1990s. Then there are those who specialize in live arranging, who need to deliver parts that can be performed every day for perhaps 20 years, and which can be played with subtly different instrumental line-ups and still work. Similarly producers sometimes specialize as live sound producer/engineers or studio producer/engineers, where a creative application of art and manipulation of sound is more achievable.

We need great producers and we need great arrangers and orchestrators, but more importantly, we need them to be respectful of each other and aware of how their various roles have adapted and evolved. Far too many producers and arrangers are fearful, skeptical, and even dismissive of each other's skill base. But producers and arrangers are essentially the same person, divided by technical specifics, history, and interpretation. The role they play and the result they achieve is almost identical, but the processes are different. Those who seek to rationalize production as the child of technology and portray arrangers as simply instrumental specialists are missing the point completely.

Just how much does *arrangement* fall under the producer's jurisdiction? If you are thinking of arrangement in a traditional sense which concerns notation, scores, stave, and dots, then how involved the producer may get depends on his or her own skills or those of the artist or band they are working with at the time, and, of course, the requirements of the track. If we consider the role of George Martin with the Beatles, his compositional, arrangement, and scoring abilities were used extensively.

However, it is not always a necessity for a producer to read and write traditional notation and if a project does require this skill then a specialist arranger can be called in to arrange a string section, for example. Modern production and producers could be considered as direct descendants of the traditional arrangement and arrangers. Although a bold statement to make many modern producers could be considered as the new breed of arrangers but simply use technology in order to realize the end result. In a similar way the modern mix engineer could be compared to that of an orchestral conductor. Mixing itself isn't necessarily a new idea as conductors have been doing it for quite some time, albeit in a slightly different context. Whether you agree or disagree with these ideas they are worth considering.

So much of arrangement is about understanding the structure of music. With a song it is very much about understanding how the various components of the song work together. This means that although a producer may not arrange a track in the traditional arranger sense she will influence the arrangement via alterations to the structure of the track, which instruments play where and when, and how they play certain parts. These things in turn alter the way in which instruments interact and go toward creating a groove or vibe.

Modern DAW software has enabled this to take place to an increasing extent in recent years and it is this technology that has allowed producers to discover new ways of piecing music and instrumental parts together, taking them apart and then putting it all back together again in a totally different way (remixing!). If this is not arrangement in the traditional sense of the word, then what is it? Production? In some respects production could be considered as the orderly management of sound and recording if traditional arrangement is considered as the orderly management of music/instrumentation.

ARRANGER TO PRODUCER: AN EVOLUTION?— BY BRIAN MORRELL

In some ways producers are essentially evolved and devolved arrangers. Production is an evolution of the art, mindset, and approach that arrangers gave us: the desire to change, evolve, restructure, develop, adapt, and arrange. This is production in a nutshell. The mindset of the arranger is therefore reborn into a modern and technically savvy production context and environment.

Production also constitutes a devolution of arranging because it essentially devolves the creative organization of sound and music into two wings: those who understand voicing, orchestration, instrumentation, and the orderly and creative dividing of instrumental textures, and those who possess a greater knowledge of how to manipulate sounds within a recorded environment in order to create new colors, textures, possibilities, and dynamics.

Some people rationalize production as if it is some great new experience and process; in fact production has always been the prism through which music is heard and appreciated. The producers of the 19th century were the great romantic composers; they were composers, arrangers, and orchestrators before such roles and responsibilities were devolved and fragmented in recognition of their own individuality, a more tech-savvy environment and a more diverse music industry. The music of Debussy, for example, was as much informed by a vast understanding of sound, sonic management, and orchestration as it was by his distinct and abstract harmonies. He was perhaps the greatest innovator and producer of his generation. In the early 20th century, long before production skills and technological advancements allowed for such sonically advanced recordings, the arrangers would act as producers by embedding the dynamics and subtleties into the arrangement. The arrangement was the mix; the recording was simply a crude capturing of the arrangement.

As an example of the power production wields, most composers nowadays do not arrange or orchestrate—something that would have been unthinkable even 100 years ago, when such skills were part of the process of achieving popular music. Most do not conceive their music whole of its instrumental sound but instead tend to think about such issues as being distinct, separate, and subsequent to the art of composing.

But, tellingly, most composers and songwriters do think in terms of production, so perhaps this represents merely a subtle shift in emphasis; perhaps production is simply a modern cool prism through which songwriters rationalize their music. Production, therefore, enjoys a massive arc of responsibility and power; it is both the arbiter of musical judgment and the prism through which most people listen and rationalize music. Again, in this respect it is identical to the role of the arranger and orchestrators.

Arranging a piece of music means much more than simply moving blocks of audio or MIDI on a sequencer or DAW around until it works best for the song. This might work for some musical genres and pieces, but it is the true arranger who will consider the rhythm, harmony, and melody of a piece and manipulate, massage, and reorder these to achieve a successful outcome.

As Brian Morrell has eluded to in his accompanying sidebars, arrangement is an art form and something that takes years to master. In this context, we are referring to the arranger that can take a lead line and perhaps some other aspects of a piece and completely reinterpret it to realize a new piece. If one was to be blunt about this in the way in which we've painted it here, the modern equivalent might be the remixer in today's popular music. The remixer will take pertinent and prominent parts of the main track such as the vocal and the main riff perhaps and envelop this within a completely new arrangement with new instrumentation, and so on.

As such, the skills required by the traditional arranger, say one who worked with big bands reinterpreting classic standards would be well-versed in instrumentation, ranges, and rhythm, as well as melody and harmony.

RHYTHM

Rhythm is one key ingredient that can be manipulated to achieve a complete difference in the arrangement of a piece. Understanding the traditional arranger's skill in managing rhythm and how to reinterpret music is key to modern production skills also. Taking a Gershwin piece and reinterpreting it will provide a key opportunity to develop or alter the rhythmic sound of the piece, giving the track a whole new life which will excite some and perhaps turn off some traditionalists.

However, what is rhythm in this context? Is it simply the tempo? Is it simply the time signatures and the relationship between the two? Of course there are so many more aspects to rhythm that we need to consider. In the scope of popular music, groove is an important aspect to the music and will be a hook to many. The hook can be a nonmelodic aspect of the song and can be a loop which consists of drums, bass, and perhaps keys. Robert Orton says he looks for all types of hooks as he begins a mix.

Rhythm in this context is much more than the setting of the piece in terms of tempo and time signature. It is interrelationships between the rhythm section, what they play, how they play it, and how strong it can be in delivering and translating the music to the listener.

Much of popular music production is based on the groove, rhythm, and feel of the music. As such, many of the producers we have spoken to in the preparation for this book have made important comments relating to getting the drums right: setting the vibe and without a good kit and drummer—go home.

Getting the sound right and capturing it is one aspect. Performance is critical. A producer should engage in this ingredient as it can be argued that the rhythm is the foundation on which the typical pop track is built. Taking time to investigate methods, tricks, and tips to improve and develop this foundation will provide concrete support for building up the track.

HARMONY AND MELODY

Alongside rhythm, the melody and harmony make up the whole building structure musically. These elements are the fabric that dictate much of the impression of the track. The rhythm, harmony and melody detail the music to us in forms we understand. We hum the tune, we are moved by the modulations, or the flow of the harmony.

It would therefore be rather remiss of the modern music producer to not develop their skills in this area. Many producers come from a classical tradition and are well-versed in harmonic theory and musical structure. Some producers (perhaps the engineer producers) come from a different perspective insomuch as they can hear what they want and may not so immediately be in a position to communicate this to session musicians in their language. Whichever camp you come from, both have merits. An engineer producer will be driven by instinct, emotions, and feel for the music. This may be in the context of engineering, and as such trials of melody and harmony might be developed in conjunction with an effect or technique within the studio.

In either camp, the producer will be required to realize the artists' needs in the production, taking ideas and assisting in the musical development of the track. Of course, how the band and producer arrive at this outcome is all dependent on that particular session. To predict how to manage each situation is somewhat naïve as each will be different and will naturally depend on the music at hand, the studio, the team, and of course the talent and creativity of the musicians.

Developing melody is key to the creation of many a hook in popular music. Being able to identify and develop the hook in the material can often be the thing that a producer can seek. Developing the thing that the music-buying public will engage in is again of paramount importance. However, there are often challenges in changing or developing a melody. This is often seen by many as the most precious of coveted aspects of a song. Some musicians will not be happy with the suggestion of a change, however small. Many, of course, will be happy to try new ideas to see how they impress upon the song. Below, Danny Cope talks about melody, why it is important and the things you should consider as a producer.

MELODIC ANALYSIS—BY DANNY COPE

Danny Cope is currently the Course Leader for the BA (Hons) Popular Music studies degree program at Leeds College of Music in West Yorkshire, England (the largest specialist Music College in the U.K.). For the past seven years, he has lectured in songwriting, song production, and popular music performance. He has also worked as a session bass player for the past 13 years. Danny has a publishing contract with Daybreak Music Ltd. in the U.K., has released five solo albums, works as a songwriting consultant to the Open University, and has delivered songwriting seminars around the U.K. In addition to working as a writer, player, and educator, he has also written and presented a tutorial DVD entitled *Everything You Need to Know about Setting Up a Bedroom Studio*.

Ask yourself which element of a song you are most likely to find yourself carrying around in your head. I'm guessing it will be the melody. Unless you possess the enviable skill of being able to sing polyphonically, you won't find yourself humming or whistling a chord sequence, and I am yet to hear anyone spontaneously burst into the recital of a set of song lyrics unaccompanied by a melodic line. It's the melody that we carry around with us. That's what we involuntarily find ourselves whistling or humming from time to time, and it's the element of a song that is most easily transportable in that sense.

If we take on board the fact that a melody is the element of the song that the listener is most likely to transport around with them, then it makes sense to ensure that the melody does all it can to ensure it can be safely carried on its way. There are three things the producer can focus on to ensure that this happens. Firstly, there's the issue

(Continued)

of simplicity. There's nothing wrong with simplicity in a melodic line. Nothing at all. I like to use the analogy of games to make this point. The best games are those that have simple straightforward rules that facilitate the playing of the game in no time at all, but that remain entertaining and engaging to get into. How annoying is it when you set out to play a board game, and you have to study pages of instructions to have a clue how to start? There are times when the complicated nature of the game is the appeal, sure, and this reality does translate into musical terms too.

It is true that, on occasion, the involved nature of a composition contributes to the fabric of its appeal, but if we're honest, the enjoyment of a melody in a pop song is rarely dependent on the listener appreciating the academic merits of its construction. The listener is, and therefore the producer should be, more concerned with how easy the melody is to "get into" and how it makes them feel. The more straightforward a melody is to get into, the more likely the listener will get involved.

This "getting involved" leads into the next point. A good melody should be singable. For the producer, specific pointers on this one relate to issues regarding range and breathing. First, be aware that if a melody is being written on an instrument, consideration of breathing might not play the integral role it should. There are undoubted benefits to writing melodies on instruments other than the voice from time to time, and if you haven't tried it, you really should. However, when you do, ensure that you test the ideas on a voice too. A melodic line will be no use to anyone if the vocalist keels over after the opening few bars.

Second, the range of the melodic line should take account of the average range of not only the vocalist, but the listener, too. If we recognize that we want our melodies to be transportable, then making them singable is vital. Obviously, different people will have different vocal ranges, and it's not realistic to ensure that every melody we write is easily singable by everyone on the planet, but there are a few things that we can and should think about. Safe advice tends to be that the range of the vocal should ideally not exceed the interval of a 10th if we want it to fall comfortably within the limits of the average vocal range. In addition to this, it's sensible to ensure that the upper limit of the range isn't being pounded! In other words, consider what the top of the range of the vocal line will be, and save it for key moments that will act as the highlight (literally) of the melodic line. As with most things in the assessment of songwriting, there will always be the exception to the rule, and it is true that the appeal of some melodic lines is founded on the fact that the range is fairly extraordinary.

Mariah Carey has established a reputation for having seemingly limitless bounds in her range, and it is undoubtedly a selling point for her records, but the fact remains that if we want people to sing (not squeal) along with our songs, we need to think about keeping them in the tonal ball park where people can, and hopefully will join in.

Finally, we need to think about making melodies memorable. It's not easy to state what will make a melody memorable at all, let alone in a paragraph, but again, there are pointers that should help. Remember that a melody is built on rhythmic and tonal variation. We can and should use that to our advantage. A monophonic melody can be memorable if it is conveyed in an interesting rhythm. Similarly, a simple rhythm can carry a great melodic line if the tonal variation is well-considered. As far as rhythm is concerned, think about the variation of note lengths from section to section and within phrases. Also think about the employment of space in the melodic line. A melody that lasts for eight bars doesn't have to fill every beat within those bars. Sometimes, it will be far more effective to leave a percentage of it free, especially when the pattern of space in relation to activity complements a different pattern in another section. In tonal terms, think about the variety of pitch distance from each note to the next. For example, do any notes repeat on the same pitch? When and where do you step up or down in the melodic line, and how big should these steps be? Generally speaking a good balance of notes that repeat, that move up or down by one pitch at a time, and that move up or down by a tonal leap of some distance will lead to an interesting melodic line with shape and character. Leaps, used sparingly, will often carry key moments and are often effective when used as part of a hook line, for example.

In all of these issues, don't forget that repetition can and should be employed carefully and thoughtfully. Study some of your favorite songs on melodic grounds and purposefully look out for how melodic information has been repeated. You may be surprised to learn just how few ideas have been strung together to create such a great overall impression.

Skills in melodic development are important, whether instinctual or learned. One may develop these skills not only through understanding from many years experience of listening to music and working with artists, but also via one's understanding of musical theory. Whereas some people will rely solely on their ear and appreciation for what makes them move musically, others will develop a tune based on their understanding of the theory and harmony attached to the melody.

Developing these skills can be interesting. For example, there are producers who do not comprehend music theory in the same way a professional session musician or contemporary classical composer might, but their ability to develop the music is unique and fresh, and sells records. The key is whether they can communicate their ideas to the musicians and technical team they are working with.

Performance

A great arrangement, high quality flexible capture, and an excellent standard of musicianship equals a perfect take. Or does it? Why is it that even when everything is in place the perfect take, at times, can seem so elusive? And what is a perfect take anyway? The judgment of this question is often left to the producer to make. The performer of course will have an opinion, which needs to be considered, but as producer you should possess the ability to remain objective and clearly focused on the end product. Knowing when you've got the best performance is often a gut feeling; however, recognizing signs of performer fatigue and realizing when you've got what you need are skills that should be developed.

Enabling the artist to produce their best performance on cue is an art, and the psychology of performance is a complex subject, which we will not attempt to delve into too deeply here. However, it is worthwhile to consider the various factors that are at play when working with artists in the studio and therefore develop a greater understanding of what may be required of the producer's role.

As we have previously alluded to, the producer's role becomes a balancing act between enabling creativity and encouraging productivity. The pressure to produce a great performance within a certain timeframe is not always conducive to a creative environment, and the producer can quickly become the catalyst that will make or break a session.

So how can a producer capture that perfect take? This is indeed the golden question to which there are a multitude of answers, some of which may not be related to the music at all. At an elementary level the producer must use his or her common sense. Making sure the artist or musicians are comfortable in the recording environment is not rocket science and changes should be made to ensure this is the case. Basic issues such as lighting, temperature, and positioning can all have a marked effect on the performance given.

On a more technical level, headphone mixes are crucial to a musician's well-being and this is where the communication between performer, producer, and recording engineer is important. Many an experienced producer or engineer will recognize the need to spend time on getting this right. Once these basic yet important issues have been addressed, the performance envisaged by the producer may be relatively easy to capture. However, as many of us would acknowledge, this is not always the case.

At this point it is the producer's ability to communicate and have empathy with an artist or musician that can really make that difference. Being able to understand their artistic vision and marry this to your chosen production watermark is key to success, and it is from this relationship that a great producer–artist collaboration can occur. These interpersonal skills and traits are things that should not be overlooked and can only be developed with experience over time. If only it were possible to teach these qualities, many of the issues our student producers come up against would be easily solved! It should be noted that in some situations the problem may not require a musically related solution at all. Being able to talk, discuss, and listen to life related issues might be all that's needed to put a performer's mind at ease and therefore enable them to give their best.

Remember, most people listen to music for an emotional experience (Levitin, 2008); it is therefore essential that the performer be able to inject emotion into their musical performance in order that this might be communicated to and received by the listener. Above all, the producer should be aware of the ingredients a session and an artist need in order for this emotional delivery to take place. In his book *This Is Your Brain on Music* (Dutton Adult, 2006), Daniel Levitin discusses the emotive recordings of the great 20th-century pianist Arthur Rubinstein. He quotes a music critic, "Rubenstein makes mistakes on some of his records, but I'll take those interpretations that are filled with passion over the 22-year-old technical wizard who can play the notes but can't convey the meaning." This is, of course, not to say that technical accuracy and proficiency are not important, but that it is emotion, passion, sentiment, and *feeling* that a listener requires in a performance in order to have a truly musical experience. As the producer it is your job to coax this performance out of the artist by whatever means possible (within reason!).

The differences of working with a group and a solo artist should also be mentioned. Although the same skills are required to capture a good performance, the ability to produce the group's *sound* and not that of a collection of individuals is important. Roy Thomas Baker came up against this situation while working with Queen in the '70s and '80s. As individuals, the members of Queen were all very capable of writing material and it was therefore necessary to harness and encourage a sense of collective creativity that would result in the Queen sound we know today. In this way many producers could be considered as sculptors working to mold a band's sound. Thus, the final product becomes the blend between the band's artistic creation and the producer's watermark.

In conclusion, the final performance that is required can be influenced not only by producer and artist but also by A&R and labels, who have a vested interest in the commercial viability of the material being produced. How much the end product is altered due to this factor is variable, and this is something that should be considered within your triangle of influence at the commencement of a project.

THE IMPORTANCE OF STUDIO PERFORMANCE

Performing in a recording studio is a very different experience for a band whose time has been spent in the rehearsal room or on stage. The dynamics are different and the feeling that you are being watched is always there. One colleague describes the studio as being a "microscope: unforgiving and critically painful for the performer."

To the inexperienced performer, the studio may seem daunting and intimidating. The session will see a performance of the musician's material being recorded and instantly analyzed for mistakes, song structures that could be changed, and so on. This exposure can feel as though the material is being ripped away from them and is being manipulated in a large record company machine.

It is therefore the producer's role to ensure that the artist is briefed prior to going into the studio on exactly what will happen and how to deal with suggestions when they are made. There has to be the opportunity for mature dialog between artist and producer to gain consensus. During planning, it may be worth considering some contingency for explanation and consensus seeking.

An opportunity to work with the band in pre-production at their rehearsal space can be important to see how the performance can pull off live in a familiar environment. Whatever attention to detail can be worked on here can be of great use later on.

It is common in live performed music to begin by recording the drums and bass. This forms the backbone to the music that is laid on it. It is very important to get these elements right and tight. The need to ensure that the rhythm section is locked together ensures groove and also cohesion in the music. Should you ever hear music, whether it is a live performance or a recording of one, where the bass is not in time with the drums, or the odd note is dropped, your attention is either drawn to that note, or the focus altered of the presentation.

How do we ensure this cohesion? Within the pre-production stage it is imperative that the band are well-rehearsed and that any changes to the music happen in enough time for a full rehearsal to take place, thus saving time in the studio. This offers the producer the opportunity to iron out any mistakes or issues that the studio will pick out, and also to overcome any performance weaknesses. However, being realistic, this is unlikely to always be the case and quite often the session will be a mixture of rehearsal, exploration, and recording, but this should be evident during initial discussions with the artist and label.

Every text in music technology speaks of the importance of headphone mixes for the musicians. Yes, we agree, but what can be done to improve the experience? There are many tricks to the way in which the mix might assist the artist's performance. Most musicians like the idea of the cans (headphones) being loud and offering them a feel that they might experience from wedge monitors. The traditional Beyer DT-100 headphones do not always offer the deep and bass-filled sound usually attributed to a live performance. So the choice of headphones can make a difference. However, there are other unconventional strategies available.

Many musicians really can benefit from being in the control room, so long as line of sight can be maintained. The guitar and bass can, for example, play in the control room with the speakers at an appropriate level to ensure essential feel. This can be good, but some musicians are used to hearing their instrument the loudest, as they stand in front of their substantial wall of amplifiers. To recreate this, a headphone mix can be created to reflect this for that musician, but also it is possible to use the stereo spectrum of the headphones differently. Balance the audio mostly to one side and place the guitarist in the same room as his amps. The quieter side of the headphones can either be placed behind the ear leaving it exposed to the amplifier, or can be left on to attenuate it slightly. Not all musicians like this way of working, but it can be a good strategy to allow them to mix and balance their headphone mix ever so slightly from their position.

The headphone mix should be as good as the mix you are hearing, if not better. It is always sensible for the engineer to solo the auxiliary master so that a mix can be estimated on the control room speakers. All good engineers will have access to a pair of headphones connected to the same headphone circuit for final checks. This mix should be as solid as it can be. Some musicians will need their presence to be louder in their mix. It is natural that the drummer will require the bassist to be louder, and vice versa, than the singer's headphone mix. Additionally, the use of effects can be very comforting to singers in their mix.

To add some reverb to the mix, perhaps louder than normal, might improve the way in which they feel they are hearing themselves. Not all people like to hear their own voices, not even all singers!

Ensuring that the musicians have the right to ask for a change of headphone mix is important. The engineer should ensure that the mix they have is what they need and will allow them to perform to their best. However, there are occasions where engineers bluff the musicians. Some engineers have been known to say that they have addressed the headphone mix as requested, but to have done little or nothing at all! Engineers may employ this bluff to overcome an over needy musician requesting frequent changes that require his instrument louder and louder to the point of spilling onto a nearby microphone.

Lines of sight between band members are again important to ensure the performance cohesion required. Setting the band up in such a way that they can see each other and work together is important. To ensure that this is possible and to minimize spill is a challenge and the permutations will be limited to the equipment and space available. Sometimes it is simply not possible and the best should be made of the situation. There is one studio used by the authors which has no window between the live room and the control room. Many recordings have taken place where the control room can only see the musician perform via CCTV. On many occasions this has been a positive thing, as the performer feels that fewer people are watching, as there is no window there. This can often allow people to let go. However, line of sight between musicians can be maintained within the live room.

Capture

A largely bandied around statement these days is "we'll fix it in the mix." What exactly will be fixed in the mix? Granted, there are many things today that can be manipulated and managed, such as editing. However, it is not always possible to edit everything due to spill, the drums and cymbals played at the edit points being different, and so on. Additionally, you cannot always tune everything you need to.

The same doubt can apply to whether a poorly captured instrument can be fixed in the mix. There are indeed many differing tools that can alter and help the mix. In some ways this aspect is easier to fix than others. Nevertheless, the need to capture an excellent recording from the beginning is vital.

During the visioning of the project, the producer will have an idea of the sound that is required. Is this band supposed to sound like a late '60s Soul/R'n'B outfit recorded with today's equipment, or are they supposed to be recorded using the latest technology and be larger than life? These considerations will dictate the equipment used and the method by which you record.

Should the attempt at the '60s recording be authentic, the choice of microphones, equipment, placement, and so on, will be very specific. The amount of technical input the producer will have is dependent on their areas of expertise and the relationship they have with a recording engineer. Looking back at bygone eras, the producer's role in the art of capturing the performance was partly forged by the increasingly mentioned George Martin. His ideas, along with engineers Norman Smith, Ken Scott, Malcolm Addey, Ken Townsend, Geoff Emerick, and so on, helped to create and capture the *Beatles sound*. There has to be a high level of communication and understanding between producer and engineers and both need to be able to understand the vision with regards to the ethos of the project. In the case of the Beatles production team, this required a willingness and understanding in being experimental and trying new techniques and ideas, from both a technical and musical perspective.

From the producer's angle much of the process of capture should be about attitude, approach, and priorities and not necessarily technical detail, such as which type, make, and model of microphone. Too many student engineers or producers place an over-emphasized importance on the gear that they "need" in order to capture a certain sound or performance. Certainly decisions in this area are important and need to be considered; however, as is said many times by a variety of textbooks and industry professionals, there are so many other links in the chain that can at times determine the quality of capture and these should not be overlooked or underestimated. Whether you record using Pro Tools, Logic, or even 2" tape, it doesn't matter if you haven't considered the myriad other issues and decisions that need to be thought through.

The logistics behind a recording session are the first elements that should be considered. As the previous section mentions, how a band is used to performing live will affect how they cope with the somewhat unnatural studio environment. Whether you record "live" or overdub, where people need to be positioned (line of sight) and the issue of using headphones all need to be carefully thought through and organized.

Just expecting an artist or band to turn up and track a song without considering these aspects is potentially production suicide and will most likely end in a wasted session. As the producer, it is your job to make sure that these things are thought of and ironed out before the recording sessions commence, hence the term *pre*-production. This is where being organized is important; discussions need to be had with the band or artist (first and foremost), with the studio, the recording engineers, session musicians, even the record label. Everyone needs to be aware of the requirements of the project and someone has to ensure that this comes together at the right point in time. That someone is the producer. So an organized, open, and flexible approach and attitude can be added to the checklist for a quality capture.

The essence of capture is about the photographic snapshot of a performance and how it is represented. As with photography, light, lenses, and illusions can make a picture leap out of the page. In the same way, a musical performance can be represented in a powerful and delightful way which leaps out of the speakers, or is purposefully integrated with the band, locked in musical synergy. In either way, capture in this sense is not only about looking at the sonic qualities but also the vibe and the mood of the music at the time of performance.

Having said all that, the equipment is, of course, very important and without it we'd not really get very far, so please forgive our statements if you got the impression the equipment is not that important. The meaning of our message is that, in this day and age, we're very fortunate in that the equipment is of such high quality that it is theoretically possible to produce Grammy award-winning stuff on your laptop (perhaps even on an iPad?). We'd like to stress again that we say "possible," because it is the knowledge and experience that produces and records great tracks more often than not.

EQUIPMENT

Most engineers love gear. The pictures of the plush recording studio can be so appealing, with the long mixing console which to a young person resembles a flight deck of the musical Starship Enterprise. This dedication to equipment and what it can do is something that we continue throughout our engineering lives and is something that intrigues us.

As audio engineers and producers, we naturally strive to improve the sound of things as we go forward and therefore our keen interest in equipment should remain, but must not override the act we're supporting: the music. All too often people hide behind the equipment rather than good music and we hope that the comments we gleaned from all the research and interviews for this book show that the music and the artist are paramount to the outcome. The gear simply supports.

Think of it another way. Many people are on record saying that 4-track made you make decisions and commit to the way things would sound much earlier in the process. Recently there have been rumors of a mix engineer receiving a track with three different drum recordings for the same track (each with 15 tracks of close mic!). This has been left for the mix engineer to decide, not the producer, which seems a little odd.

Can it be assumed that restrictions can produce serendipitous results and great music can be made on high-quality 4-tracks with a high-quality studio and staff around it? Many bands have tried it and it does suggest that a well-written song, with intricate arrangements and imaginative engineering, can produce albums such as the Beatles' *Revolver* and *Sgt. Pepper's Lonely Hearts Club Band*.

There are other examples where a much higher level of technology has come to bear unique musical outputs, such as on Def Leppard's 1983 album *Pyromania*—a seminal album of the time produced by Robert John "Mutt" Lange. However, disastrously, the band's drummer Rick Allen lost his left arm in a car crash toward the end of 1984. Wishing to support Allen's return, the band encouraged him to work with the electronic drum company Simmons to develop a system in which Allen could use his feet to play combinations of sounds which would allow him to perform live and record the forthcoming *Hysteria* album. This advance in technology not only allowed Allen to return to the band and continue his career, but because of the time spent in the making of the album, it is reputed that Lange spent many hours ensuring every part was played to perfection, sculpting out what has become a seminal '80s rock album.

Let's consider the equipment in a little more detail now. Microphones are that point in the chain where things are captured and tracked. Thus, it is often considered that these are the most important items in the room. Naturally they are important and can truly offer some considerable quality enhancements if selected correctly, and put in the right place.

Geoff Emerick, as quoted in Mark Cunningham's book *Good Vibrations – A History of Record Production* (Sanctuary Publishing, 1999), details the first session in which Emerick was placed in charge of the controls of a Beatles session:

> "I was listening to some American records that impressed and I didn't know how they got those sounds. But I tried to change the miking technique that I was taught at Abbey Road, thinking that was what it took to achieve a certain sound. I started moving a lot closer with the mics and we started taking the front skin off the bass drum. There was a rule here that you couldn't place the mic closer than eighteen inches from the bass drum because the air pressure would damage the diaphragm. So I had to get a letter from the management which gave me permission to go in closer with the mics on Beatles sessions. I then went about completely changing the miking techniques and began to over-compress and limited things heavily. *Revolver* was the first time we put drums through Fairchild Limiters and that was just one example of the things that the other Abbey Road engineers used to hate because they had done it a certain way for so many years… so why change it? But the Beatles were screaming out for change. They didn't want the piano to sound like a piano any more or a guitar to sound like a guitar. I just had to screw around with what we had."

Close miking on drums was employed forever and is commonplace these days. Mic placement is important.

Using the best mic for the job is a lot about experience. You really need to have tried a mic out in a specific position on the instrument or voice concerned before you can really ascertain its sonic qualities. As ever, there are a huge number of variables and influential factors on the end result; however, it is surprising how many aspiring engineers and producers make decisions based on what they have read, heard, or think. The honest truth is that you do not know until you try something out (consider Emerick as mentioned above). There are certainly commonsense decisions that can be made based on sound technical knowledge, but after this it is about selecting the equipment (in this case a mic) based on real-world results and experience. Knowing the sonic qualities of the equipment at your disposal is a little like a photographer knowing which lens to use. His basic technical knowledge will tell him that he needs to use a wide angle lens if he wants to capture an expansive landscape shot, but it is his past experience that will help him decide on which specific wide angle lens to select and where to focus it.

There is, of course, much more to recording and capture in terms of the equipment used and, although the mic might be the initial (and therefore the most influential) piece of equipment in the chain, there is much to follow! We do not want to list every piece of studio gear in this chain as it has the potential to be considerably long. However, it is worth taking a moment to consider (possibly with your recording engineer if you have one working with you) how the choice of other components will influence your capture and therefore the desired outcome of a track.

REPLACING SOUNDS

Another approach to capture that is common in modern production is one that results in the replacement of the acoustic sounds with samples or with the layering of samples over the acoustically captured instruments. This production technique raises some interesting issues with regard to how instruments are, or need to be, captured.

A good example of this process is beat replacement, where the acoustic kick, snare, hi-hats, and so on, are replaced or at the very least enhanced with a sampled equivalent. In this situation the priorities involved in the capture of the drummer's performance change. The capture becomes more about the recording and storing of the rhythmical timing and performance of the drummer rather than the tonal qualities of the kit itself. The rhythmical essence of the performance is what is wanted and not the acoustic output.

The audio will essentially be used as a MIDI trigger so rhythmical timing, feel, and groove are the priority, not how big the snare sounds. This presents a strange requirement to the recording engineer, who would more traditionally be used to taking great care and time over the sonic qualities and capture of such an instrument. It might take a while for an engineer to come to terms with the fact that, as long as the mics reliably capture the rhythmic hits of the drums, the actual captured sound can be ignored to a certain extent; after all, it's only going to be replaced with another kit! Of course, if the intention is to enhance or complement the acoustic kit with samples then the acoustic capture still needs to maintain quality, albeit in the knowledge that the samples will strengthen or change the final result.

There are many recordings in which this production technique has been used, but an interesting example to consider is the 2006 Grammy Award and Ivor Novello Award-winning track "Rehab" by Amy Winehouse, produced by Mark Ronson and mixed by Tom Elmhirst. This track uses very minimal miking on the original drum kit (mono, one overhead) and was recorded live alongside other instruments in the rhythm section. This was all part of Ronson and Winehouse's original vision for the sound of the track with the Dap Kings being used for their specialist late '60s R'n'B/soul sound. However, the track utilizes kick and snare samples layered over the original in order to enhance the sound and give the track a much more up-to-date commercial and radio-friendly sound. Without these samples being added by the mix engineer (Elmhirst) the track would have taken on a very different flavor and arguably not have faired so well with the record-buying public at the time. They add the drive and punch that is expected of today's records to the retro R'n'B/soul groove, something that wouldn't have been provided by the original recording.

THE STUDIO
The vibe

The choice of studio for the producer can be varied (depending on budget) and this is probably one of the first points on the quality capture checklist: where? Most producers would agree that three key ingredients generally spring to mind when considering a studio: equipment, acoustics, and feel (vibe). Of course the fourth consideration (and possibly the most important!) is whether these elements match up with the requirements of the project and artist.

In terms of the studio environment, the vibe or feel of a facility can be the make or break of a producer's choice. Having a space where the artist and performers (and producer and engineer) feel at ease, comfortable, and relaxed is of paramount importance if a strong performance and recording is to be created. Does it have natural light? Is there a lounge? Is there a pool table? Is there a kitchen? Is there a view of the outside world? Is it near somewhere to eat? Although some of these points may seem superficial at first, all of them can play a huge part in the general perception of a studio and can ultimately help create the right conditions for a successful project. The artist and band need to see the studio as somewhat of a base, a place where they can be themselves for the duration of the tracking process.

With the above points in mind, the ability to adapt a studio to the artists' requirements is one other area that should not be overlooked. Can drapes be hung from the walls and rugs placed on the floor? Can uplighting be brought in and the live rooms' lights dimmed? Again, strange things to consider but essential in terms of making the environment work for the good of the recording process. Some studios may take to this approach easier than others. Some studios may be very hard to disguise.

Producers should be aware of selecting a studio based on the clients that have previously recorded there. It may have worked for them and achieved great results, but then that was a different project at a different time with different people. The studio can give a lot but it can't give it all. Likewise, a studio's history can be inspiring and some artists may benefit from feeding off of this historic vibe. We only have to think of Abbey Road in the U.K. in order to realize this; however, this can also add a degree of pressure to the occasion. It might be difficult in certain studios to keep thoughts of past clients and recordings at a distance, and this sense of greatness is not always a vibe that helps artists feel relaxed in the process of realizing their own artistic endeavors.

Whatever studio is finally chosen, it needs to meet the artist's needs in terms of feel and vibe and as long as the studio sounds good and is appropriate acoustically speaking (see below) then the vibe can often take precedence. After all, if the artist is happy and at ease, then this can make their performance and therefore the capture of a successful project a less troublesome task for the producer.

The sound

Hand in hand with the question of studio vibe is the *sound* a studio has, or to be more technically accurate, its acoustic properties. Obviously most professional studio facilities will have the acoustics of their various rooms "tweaked" in order to give the best possible results. However, not all studios sound the same and therefore results can vary. Different studios will also have a variety of different rooms: small, medium, large, some "live" and some "dead."

It is the blend and variety of these individual room characteristics that a studio may have to offer that can give a recording a certain sound, and this is a factor that many producers take into account when selecting a studio to record in. Often studios will have reputations based on specific sounds that they are able to achieve and capture. If a lively and large drum sound is required for the album, then specific studio facilities with big-sounding drum rooms may come into play. Likewise, if a full string section is to be used then a studio with the right sized space and live acoustic properties needs to be considered. Again, the artist and their project will dictate the type of sound required; however, it is the producer's job to select a studio that will offer the best possible opportunity to capture these specific requirements.

In his book *Producing Hit Records* (Schirmer Books, 2006), David Farinella interviews guitarist and producer Daron Malakian about the recording location of System of a Down's album *Toxicity* stating that when looking for a place to record he "turned to his record collection."

According to Farinella: "The album he pulled out that he wanted to mimic was *A Date with Elvis*, a 1986 offering from The Cramps. 'I was looking for a thicker sound for the band, and I thought if a room can make The Cramps sound thick then that was the room for us,' he explains. The Cramps album was recorded at the part of Ocean Way Studios that is now Cello Studios in Los Angeles. 'We were already ready to record a thick-sounding record, so I thought, imagine what it would do for us.'"

As we are all aware, the record industry has changed greatly in recent years with budgets and economics having an ever-increasing influence over the way in which an album is recorded. With this in mind, it is worth mentioning that many albums these days are not recorded in one studio alone and that a variety of smaller facilities may be used at different points during the tracking process.

The previously mentioned point of capturing a drum sound is a good example to use here. A producer and band may select and use a studio specifically for capturing the drums or the rhythm section due to the size and acoustic properties that it has to offer, but then switch to a different and possibly smaller facility to record the overdubs, keys, vocals, and so on. In fact, with the flexibility that recording gear now offers, many producers and artists are tracking the larger or main elements of an album in professional facilities and then retreating to their own studio setups in order to finish the project. Obviously the sound of these smaller (possibly home studio) facilities is important and any discrepancies in terms of capture quality need to be addressed. However, with a reasonable live space that offers good control of the sound source being captured there is no reason why this "mix and match" approach cannot produce excellent sounding results.

The gear

Gear, or to put it slightly more eloquently, recording equipment, can be a big deal. Certain mics, a certain desk, and a varied selection of outboard are all important points on many recording engineers' checklist. As the producer, this may be just as important to you. We say "may be" as you may not be a producer that comes from the producer-engineer stable (as identified by Burgess in his 2005 book *The Art of Music Production, Omnibus Press, 2005*), in which case it is entirely possible that you are not interested in what gear is used but more the results that are achieved.

Whatever your viewpoint, you will at least have to consider or discuss the equipment that a studio has when looking for a location to record. Of course, there can be situations where the ideal studio for the project in terms of capture (sound and vibe) doesn't necessarily have the equipment you would prefer to work with. It is in this instance that the option of bringing your own gear in or hiring it in needs to be explored with the studio. In some cases the producer might be intrinsically involved in what gear is brought in, and in other cases it may be that the producer leaves this entirely to the recording engineers. Again, much of this will depend on the producer's skills and expertise and their relationship with the engineers.

Whatever the situation is with equipment, a producer should never choose a studio based on the gear alone. Key factors are quality, variety, and (therefore) options, and of course personal preference. If these factors can be met by the studio in question then it could be a great choice if the other boxes of sound and vibe are also checked. If not, then bringing in your gear from outside should enable the quality of capture you're looking for.

The engineering, equipment, microphone choice, and positioning

Engineering, equipment, microphone choice, and positioning can all have a huge effect on the capture of an artist's performance and therefore the final outcome of the recording. If you are from the technical side of things then your involvement in these decisions may be a natural process; if not, then your ability to communicate the kind of sound you want to others is key.

If you do match Burgess's description of an engineer-producer, then the important link to make is between the gear and the sound you are trying to achieve. Many of your decisions will be based on previous experience and you will know what equipment has worked well for you in the past and the (hopefully) pleasing results you have gotten from it. However, the question should be asked, "What is the most appropriate choice for this project?" If you are an aspiring engineer-producer then your experience of different gear may be limited, but it is this question that can help you stay focused on the sound that is needed and not get carried away with exotic equipment choices that might be tempting you.

As we have mentioned previously, if you are trying to capture a retro sound then perhaps close-miking the drum kit with 9 or 10 mics in a studio's drum room is not the best route to take. In this instance consideration of how many mics, their placement and type should all spring to mind, as should the tracking method; whether live multitrack or overdubbed. In fact, you'll also want to consider the types of pre-amps you might use in order to gain the desired characteristic for the sound, and this is where knowledge and experience of different mics and outboard will come into play (enter the recording engineer for assistance). If you're looking for retro warmth and coloration, then perhaps valve based equipment needs to be experimented with. Above all, even if you have an engineering background, the role you are playing as producer should never be forgotten, as the sound of the recording is ultimately your responsibility. Therefore, don't get too bogged down in mic and pre-amp type discussions, as this is one of the reasons you have a separate recording engineer.

As a producer, if engineering, equipment, mics, and so on are not your forte, then this is where the skills of a strong communicator need to kick in. Your vision for the project and therefore the way you want it to sound need to be put into words so that others (such as the recording engineer) can understand, allowing them to cross-reference and match this to their experience and skill on the technical side of things. Establishing an engineering ethos may be the way forward here, something that can act as a blueprint for the recording sessions. Again, the final capture is your responsibility so if something isn't working it's ultimately your call to put it right.

At this point it is worth mentioning that, as the producer, you may have the luxury of choosing the engineer you work with, in which case your decision will most likely be based on a working relationship, ethos, previous experience, and specialist skills. There is much to be said with regard to the working relationship between the producer and engineer (as discussed in Section B).

Recording methodology

It would be very tempting to repeat ourselves and get lost in discussing a favorite topic—recording techniques with regard to mics, placement, outboard equipment, and so on. However, here we'd like to highlight the underlying practical approaches for recording and why these need to be considered.

Your recording methodology is obviously linked very closely with your equipment choice, in that the type of equipment you use is likely to alter the way you may record in the studio. The biggest difference is apparent between the use of analog and digital formats and DAWs. It is not our intention to enter into the long-running "analog vs. digital" debate here. However, in comparison to working with DAW software such as Pro Tools, using analog tape can make the recording process more time-consuming and has certain limitations. Larger track counts, virtual tracks, nonlinear recording and editing all make the digital world a different place to work in and obviously this affects the way in which recording takes place.

Of course, one of the main concerns will always be cost and therefore your budget may dictate the way you end up working. (Not only is analog very different to work with but also expensive.) However, the approach to recording and the methods you use should again revolve around what is appropriate for the project and the artists you're working with. Some artists and producers like the perceived warmth that analog tape can bring to a track; however, when balanced against the practical advantages that digital recording has to offer, the choice may be more straightforward.

The two main questions that are generally asked when considering recording methods are "Should it be recorded live?" or "Should parts be overdubbed?" The answer to these questions should take into account the initial vision for the product in conjunction with the pros and cons of each tracking method, as this will help dictate the methods used. For example, a live and energetic band sound will require a more natural live approach to tracking, whereas a polished, tight commercial pop track requiring high production values is more likely to utilize the overdub method of tracking. It is interesting to note that it is common within the production of commercial pop music (where much of the accompaniment is sequenced or sampled) that the musicians may never meet the vocalists or artists in a studio session due to the nature of overdubbing.

Of course, the artist and band may also have a preferred way of recording (based on their past experiences and results), and if working with a group or individual for the first time it is important to establish what works for them and perhaps, more importantly, why. Having said this, an artist or group with relatively little recording experience may not fully appreciate the pros and cons of each method in relation to what they are trying to achieve. If this is the case then as the producer your knowledge, judgment, and guidance need to come into play in order to ensure both a smooth process and a pleasing result. But be aware, suggesting a change in the way in which an artist or band records—for example, live multi-tracking or overdubs—may not be easy and, as ever, diplomacy and tact may be needed to get the message across that something isn't working.

Session times for best capture

When we say *best capture*, what do we actually mean? We're basically talking about harnessing and capturing the essential characteristics and energy in a performance that are required for the track in question. It is important to point out that when we say *energy* we are not necessarily referring to energetic performances or characteristics, but more to *emotional energy*, whatever form that may take.

Out of all the various debatable issues within music production and recording (of which there are many!) the time at which sessions should take place is possibly one of the most difficult to discuss. To try to suggest that there are specific times during the day or night that will always yield a great result would be incredibly assumptive, especially given that the process of music production is such a personal and creative pursuit and therefore highly unpredictable. In fact, it is only after consideration of the wider context of the project and the individuals involved that decisions in relation to what time your sessions occur can really be made. The key to organizing anything well is good communication and booking studio sessions is certainly no exception. After speaking to various producers for this book, we realize that creativity and good performances can't really be predicted but certainly taking some time to plan out the recording schedule with the artist and band and clearly understand each other's thoughts on when and how the sessions should occur is essential.

One of the fundamental factors here is remembering who exactly is doing the performing: the artist or musician. As Hayden Bendall reminded us when talking to him at Strongroom studios in London, the artist should really be treated as the most important person in the room. Therefore, with this in mind, you really need to consider when it is that *they* want to record, and at what time of the day or night do *they* feel at their best. It may be that an evening session would work best, as the artist has had time throughout the day to relax and then is ready to perform. However, what is perhaps the most important aspect here is the fact that they *have* been able to relax and feel ready *before* the session is due to start. If this does not happen, then an evening or late-night session may be a waste of their and your time as tiredness or stress will kick in and creativity will be stifled as a result. Taking this on board, it could be suggested that the specific time the session takes place is not necessarily the most important factor but more the condition or state of mind the artist finds themselves in on a day-to-day basis. If an artist is ready and raring to go in the morning, then why shouldn't the session happen at that time? (From our conversations with producers this seems to be a rarity!) Being flexible enough to accommodate various session times is important as the project and recording progress, and again it's all in the interests of *best capture*.

In reading the above you might be starting to think that we are portraying the artist as the be-all and end-all of the decisions you make in respect to what times sessions occur. However, while their views and needs are fundamentally important, we should point out that as the producer you need to be aware of the other musicians involved in the recording session and their abilities to produce the goods at the times being suggested. There's no point in the artist wanting a session to start at 11pm if the guitarist you have brought in to work on the track simply can't produce their best at that time of night. Here the producer's role again becomes one of an organizer and negotiator, making sure that the times chosen to record suit everyone involved as much as possible. Granted, this may often not be an easy task and will vary from project to project and artist to artist, but nonetheless being prepared to manage the issues and dynamics of the situation is a must. Of course, one of the benefits of overdubbing is that it negates the need for musicians, the band, and artist to all be together at one time, so depending on the project and recording methods used, some of this may never be a great issue. (Such as the example above.)

One topic that always crops up when discussing time is the duration of a session. Again, this can be very varied and highly inconsistent depending on the project and the individuals involved; however, the guidance for this is not rocket science and can be based on an understanding of basic human needs and an awareness of the people around you. We all know that, after a certain amount of time, most people reach a peak, and after this point results might not always be of the best quality. In terms of the studio environment, our hearing is one of the first things to suffer from fatigue and therefore long sessions should be managed with regular breaks. Although it may sound trivial, the producer also needs to keep tabs on when people within the session last ate. A tired and hungry artist, band, or musician does not make for a great take! Some artists may not want to be bound by a regimented structure of breaks for food and drink and certainly the approach can be flexible, but as the producer you need to be able to call time and make a food break happen. If not, the session might bound aimlessly along and, although the artist may have the best intentions at heart, the scale of diminishing returns dictates that the results will not remain of a consistent standard. A change of scene, some fresh air, and good food can make a surprising difference to the subsequent takes and therefore knowing the best local places to eat might be very useful information for the producer to have.

Of course, overall session length is often known to be quite considerable and it is not uncommon for some producers to work anywhere between 12 and 16 hours in one day. And this may continue for a number of days, which can take its toll. Therefore, establishing a work ethic for the project might be something that you want to consider discussing from the outset. In many cases (again from our conversations with industry folk), the amount of time a producer may work on a session will vary depending on where they are in their own career. If just starting out, longer sessions may be more commonplace than those of a seasoned producer with various credits to her name. As a younger producer you may be willing to give what it takes in order to gain the experience, which is fine and very commendable, as long as the number of hours you work in one session does not hinder the quality of what you produce. Knowing when to call it a day is quite a valuable skill to acquire, and certainly individual thresholds will vary in terms of the number of successful hours spent in the studio in any one session.

David Farinella includes many interviews with producers within his book, *Producing Hit Records—Secrets from the Studio* (2006). The comments below are taken from his conversations with David Fridmann on the topic of studio session duration.

"I just cannot see being productive, at least from a production standpoint, after 12 hours a day. I can't do more than that... I've done it, and the truth is, it wasn't very productive. The extra four hours didn't help. If anything, a lot of times what happens is you get more tired or you get more angry or you get more drunk and mistakes happen. You just can't concentrate that well, so I try and keep things down to a 12-hour day at this point, which allows for breaks and meals, and two-week periods."

From our own discussions with the producers interviewed for this book, we can conclude that a 12-hour session with breaks and meals seems to be about the average duration. Naturally this was not a unanimous view, but the majority assert that any longer is generally not that productive. However, the times at which these sessions start can vary and this is something particular to each producer and artist.

When all is said and done, it is worth remembering that there are many personal factors and situations that can contribute toward the best timing and duration of a session. Artists, producers, and engineers are all human and although some of the factors might not be that rock 'n' roll, they are real-life considerations that can influence the recording sessions and the project more than is perhaps immediately apparent.

Early on in a producer's or artist's career, personal commitments and responsibilities may be few, which on one hand can be a bonus. However, even after a session has ended, the young artist or band might be left wondering what to do and where to go. As the producer you may feel that your role extends outside of the studio and that you carry a certain amount of responsibility to ensure that the artist is comfortable and knows where to eat and relax, and so on. You may even end up spending the time outside the studio together, as this can be good for the artist–producer bond. However, as time goes by, friends, family, and leading some kind of life outside of the studio and your work might become more consuming. Therefore a balance needs to be struck in order that the time in the studio spent working can be uninterrupted by thoughts and phone calls regarding the other areas of your (or the artist's) life.

This is certainly easier to say than do and we shall not pretend for one minute that we can offer a solution as to how best to achieve this, as many of the choices are down to personal circumstances. However, although the music industry might not be the easiest career path in terms of work and life balance, as a producer it is worth establishing what is important for you and the artist and therefore establishing a way of working before the sessions commence.

THE ENVIRONMENT

In previous sections of this chapter we discuss the studio, its *vibe*, *sound*, and *gear*. Therefore, another section addressing the environment might seem like needless repetition. However, let's take a step outside of the studio for a moment and consider the *external* environment and surroundings in which a recording may take place. By getting wrapped up in choosing a studio and the gear for a recording, it is very easy to forget the place in which the studio is located and how this can have a considerable effect on people and proceedings. Although much time and consideration may be given to choosing the studio itself, how much importance and consideration should be placed on the surrounding environment? Yes, studio sessions can be long and take up most of the day (or night); however, when the session finishes it is what's on the outside of the studio that most people are interested in.

It may be that, as the producer, you will not have a great deal of choice as to which studio or which part of the country (or world) you will record in. This could simply be down to budget or possibly the logistics of getting all the necessary people to the location concerned, or even that the artist has a particular location in which they feel most comfortable. However, let's assume that you do have some choice in the matter and highlight some of the factors that should be considered.

As with many things in music production, decisions cannot really be made without the consultation of those involved, specifically the artist and band, and therefore when considering the attributes of the surrounding environment the artist's views and thoughts need to be taken into account. It may be that a band and the artist find the surroundings of a city with a thriving and energetic night-life particularly attractive, as this will provide them with a much needed escape from the relative confines of the recording studio and give them time to let off steam and relax. If this is the case, then a studio with a good city location is ideal. Think clubs, food, and entertainment. However, on the other hand, an artist might find the space, peace, and tranquility of a more rural location to be the perfect antidote to the demands of the recording process. Going for a long walk in the open air between sessions or at the end of the day can, for some, clear the mind, allow them to relax, and therefore be of real benefit to the creative process. Think views, fresh air, and countryside.

Although these examples might seem a little extreme, do not underestimate the power that the surroundings of a studio location can have on the recording process. The way you and the artist spend time outside of the studio can affect your performance and what is produced inside the studio. Remember that many factors can contribute toward the creative process. But a word of warning: What may seem like a nice idea at the time may in reality not be what is needed. For example, a rural location, while appealing in many ways, can be limiting. What if you want a McDonalds at 1am, for example?

There are many artists who chose different recording locations based on the surrounding environment and not just the studio itself. Sting is one such U.K. artist who initially springs to mind, with albums such as *All This Time* (2001) being recorded at one of his homes in Italy. Certainly this album sprang out of a slightly different set of circumstances (Sting's desire to perform a live concert from his home). However, in general he is an artist who seems to place importance on the surrounding environment in which he records and recognizes the influence it can have on the overall recording and production process.

In a short film (*One Song, One Day*) documenting the recording process of his song "In Repair," American artist John Mayer refers to his use of New York's Avatar Studios as being "part of my life, when working and living in New York." Although this choice of location is undoubtedly to do with the studio and convenience as much as anything else, Mayer's comments allude to the fact that many artists want the time spent in recording studios to fit into their daily routine.

If you are just starting out in the business of music production, then the idea of having such choice of studio location is possibly a distant luxury. However, there are alternative methods and locations to the stereotypical city studio that can be used to produce some great results (especially with the technology that is now available). Careful consideration will be needed in terms of acoustic properties and equipment logistics, but the ability to record in a physically different location (and therefore surrounding environment) might be exactly what the artist needs. Consider hiring a house in the country, for example; while this might seem expensive, remember, you will not be paying by the hour as you would in a traditional studio and you get the advantage of the rural surroundings and vibe.

From interviewing the various producers for this book, we cannot help but notice a seemingly common topic with regard to recording and the studio: food and drink! That's not to say that the professionals we had the pleasure of speaking to were all food fanatics, but more that each one seemed to mention the process of eating outside of the studio and the benefits this brought to the production and recording process. Dining out can be a very social activity and most people would agree that eating a meal with others is both a relaxing and positive way to spend their time. Therefore, when taking into account the external environment a studio is located in, don't forget to check out the local restaurants and cafés. Or if you're out of the city, consider who's going to be doing the cooking for you!

SECTION D
Doing It

The Session

CONFIDENT CREATION

The studio session can be one of the most exciting times in the creation of a project, where the fruits of pre-production can begin to be seen and where musical spontaneity can create a very special result. The recording session is the stage where everything seems to happen, certainly in a capture sense. Artists, bands, session musicians, engineers, and producers all combine under one roof to create. It is the point in the project where ideas become a reality and a "product" starts its life. The fact that the ideas and spontaneous moments become "committed to tape" at this stage can be an issue for some and this can often lead to tense and nervous environments in which to try to be creative. Being able to create and inspire confidence within the recording session is a skill that all producers need to have in one guise or another. Once established in the industry, many producers may take this as a given; however, when first starting out down the rocky road of music production, the topic of creating and inspiring confidence may be a difficult one to grapple with.

So what do we mean by "confident creation"? One of the key factors here is believing in what you are doing. There's no good in being involved with a project for which you lack enthusiasm and vision, as this will hinder your abilities to steer the project and inspire confidence in others. It is sometimes surprising how saying nothing can be the loudest form of communication and, in the relative confines of a recording studio, body language can say an awful lot as to the true thoughts and feelings of the individuals involved. This lack of enthusiasm and vision can occur for a variety of reasons, but many times it can be down to the fact that you simply do not really understand the project and therefore don't believe in it. When we say "believe" we essentially mean that you are able to see, understand, and agree with the overall artistic vision and musical intentions of the project. This is where the marriage of artist, project, and producer comes into play and the involvement of the right people for the job is of fundamental importance. Knowing when to accept the invitation to be involved with a recording or not can be a tricky call to make; any doubts need to be ironed out before commencing. If not, you may find yourself in a very uncomfortable position further down the line when the stakes are considerably higher and you're performing less confidently.

What is Music Production. DOI: 10.1016/B978-0-240-81126-0.00010-X

PRODUCER CONFIDENCE

When interviewed for this book, producers and mix engineers Phil Harding and Tim Speight both commented on the commitment and confidence shown by Pete Waterman with regard to the work and projects coming out of PWL in the 1980s. His belief in the music being produced and the products being created was so strong that he was able to put himself out there and say "this is what we're doing and this is what we can do." Having this confidence worked in a positive way for the production/writing team of Stock, Aitken, and Waterman and the mix engineers around them such as Phil Harding. They quickly earned the nickname The Hit Factory due to their dominance at the top of the pop charts in the U.K. during the 1980s with artists such as Kylie Minogue, Rick Astley, and Jason Donovan. Waterman even created the phrase, "*Tomorrow's sound, today's technology.*"

Of course we're not saying that in order to be successful you *must* display a brash and outspoken confidence; however, if you find it difficult to promote yourself and your ideas, then the role of producer is possibly not for you.

Confidence is discussed in Section B.

Your influence needs to be exuded not only to the artist and musicians involved but also to the artist's management and A&R. It is worth bearing in mind that the artist's management and A&R from the label are placing a considerable amount of trust in you and your abilities, and in some cases are entrusting their latest talent into your guiding hands. It is understandable that this can create tension, especially when time, effort, and money are being invested. Therefore, being confident in what you are doing and the decisions you make is important, as many in the process will be looking toward you for a reassuring lead.

Your role?

The producer's role in the session can, as ever, be one of many things (as we will discuss later in this chapter). The role has been described in a number of ways before but we tend to think of a chameleon and the way in which it changes its colors in order to blend in with its surroundings. Likewise the producer's role in a session is one that will need to change and adapt in reaction to what is required in order to achieve a successful recording. The need for adaptability has been a common thread in conversations we have had with various producers while writing this book, and it is obvious that different individuals work in differing ways depending on their background, skills and experience. However, there are a number of generic points that seem pertinent to remember when considering the producer's role within the recording session.

Firstly, the need to maintain an objective perspective in the studio is of paramount importance. It is the producer's job not to become too bogged down in the fine detail or emotion but to maintain the vision of the project as a whole. It is the producer's responsibility to know when there is some mileage in an idea or when time is possibly being wasted. Secondly, the artist is the most important person in the room. This is hopefully common sense to most that will read this; however, it is worth acknowledging that once wrapped up in the moment of a session it can become easy to forget the reason why you are there and what it is you are trying to achieve and why. It is the artist's and musician's session. They are there to perform and record; certainly you will have your views and opinions (that is why you have been hired, after all), but your interaction should always be for the good of the artist.

Producer, arranger, and engineer Haydn Bendell describes the role as one of "a service provider" where the artist/musician should always be made to feel as comfortable and safe as possible. When discussing the importance of the artist, he gives the example of two records he may produce using exactly the same studio, people, equipment, and techniques. Yet while one record could go on to win various awards, another could be far less successful. The variable factor, the artist, is surely the most important cog in the machine. An idea worth considering when examining the way in which we all work, perhaps?

Creative

There are many ways in which creativity can be inspired and achieved and we know that many producers have a variety of styles and methods (even if they themselves struggle to explain exactly how they do it). Of course, some would argue that much of the creativity takes place before the recording sessions occur and that after pre-production there is not always time to be experimental and creative. However, let's remember that, despite the careful planning and arrangement achieved via the pre-production stage, the recording sessions are more often than not the place where ideas blossom and develop and where great performances can make a real difference to a song. The producer's role is integral in steering this creative process.

So how do you create a creative environment or engender creativity within a session? As you have by now realized, you will not find a hard and fast answer to this question, no matter who you ask (believe us, we have!). However, there are considerations and ideas that may be helpful when tackling this issue. Creativity can come from a variety of different sources, some of which are more obvious than others, but in order for a producer to create a creative environment some basic issues need to be realized.

Firstly, for any person to be creative they need to have a sense of security. Therefore, in the context of the session they need to feel comfortable with the people around them. In order for this to be the case, the people they are working with need to understand them and what makes them tick; a mutual trust needs to be developed. This may sound rather deep but a producer really needs to put some time into getting to know the artist and musicians they are working with in order for this to take place. It may sound trite, but a preliminary meeting over a coffee or beer really could help the creativity of a session further down the road.

Secondly, creativity is a very personal thing and usually comes from some kind of emotion. Therefore, creating an environment that someone can feel relaxed (or safe) enough to be emotive in is essential. An artist or musician needs to feel sure that whatever they do or say in the studio stays in the studio and that they are essentially able to be themselves. Therefore, it might be worth considering who is around the studio during recording sessions and the space that is used in order to capture a performance. If you want an intimate and personal performance and delivery, then the live space of Studio 2 at Abbey Road may not be the best place!

Thirdly, think energy! It is difficult to be creative in an atmosphere that lacks energy. There's nothing worse than trying to be creative and bounce ideas around when something or someone in the session is bringing the vibe down. By energy and vibe, we mean a sense of positivity or sense of enjoyment. Certainly we acknowledge that some great and very creative results can come from a negative situation (just consider the personal relationship issues surrounding the Fleetwood Mac *Rumours* album, for example). However, in general the recording session should be an enjoyable, positive, and creative experience. If people are not enjoying the process, then the producer needs to identify why and seek to address the issue. In many ways the initial vibe for the session is created and set by the producer, something which at a later point the artist and musicians will pick up on and feed off of. The producer really needs to have his finger on the pulse of the session.

An interesting psychological perspective and methodology to consider while discussing this topic comes from that of producer Brian Eno. *The Art of Record Production* by Richard James Burgess (2005) describes how Eno developed a set of mood cards for use when artists are seemingly at a creative dead-end. These cards pose statements designed to provoke a response of some kind in order to restart or reroute the creative process.

Allowing time and room to experiment can be a very liberating and fruitful approach for both the artist and producer. Many anecdotes can be told by seasoned professionals whereby experimentation and serendipity has led to a pleasing result. However, creativity has inherent dangers if left to run wild, and this is where the producer needs to have control over the session and be able to discern a good idea from a bad one. Being able to perceive and judge how a situation is unfolding and what the possible result will be is an essential skill to have. This is where industry names such as Brian Eno and Trevor Horn have received much praise from artists they have worked with. Not all experimentation leads to the desired outcome, so the producer has a responsibility for quality control as it were, making sure that whatever is being created still fits the overall aim of the session.

This section could quite easily turn into a lesson in psychology (as could many of the sections within this book) but this is not really our aim here. Therefore, while contemplating the topic of creativity and confident creation we'd suggest considering the following four simple mindsets or approaches, which may prove useful within the studio session. We'll call them *director, catalyst, nurturer,* and *psychologist*; of course much depends on the situation you find yourself in and as such these mindsets will become interleaved as the sessions take place. Being flexible is key.

PRODUCER MINDSETS AND APPROACHES

Director

Organization, imagination, vision, decision-making, and leading by example are all qualities or skills that feature within the *director* mindset/role. As the term suggests, this involves taking on the direction of the session and those present. In some ways this could be likened to that of a film director (as the producer's role often is) where the actors are directed and instructed to deliver scenes or lines in a certain way and where technical personnel are directed in order to achieve specific results. Obviously this type of approach is neither applicable nor appropriate with some artists or bands, particularly those who are more established within the industry and therefore have more recording experience. However, some individuals and groups may thrive off of this direction and once instruction is given they have clear boundaries within which they can work and be creative. It is also worth considering that this direction may only be required in some and not all areas. Some artists may not come from a formal or traditional musical background and therefore may require specific direction and support in certain musical areas in order for them to feel more confident in what it is they are trying to do. Or it may be that their natural approach, while very creative, lacks focus, and therefore direction is needed in order to concentrate the flow of ideas. In the context of confident creation it is important to point out that the director mindset and approach should be considered as one of leadership and not dictatorship. Good leadership gives direction while allowing for individual creativity and freedom; this in turn can inspire confident creation.

Catalyst

Acting as a catalyst in a session or adopting the catalyst mindset can be a risky route to take as the outcome is not always predictable. Here the commonly used analogy of "a piece of grit is needed to make a pearl" is useful to consider. Sometimes it will take an idea, a comment, or even deliberate provocation in order to get the creative juices and performances flowing. Artists and bands can become bogged down in ideas or ways of working and sometimes an external stimulus is needed to change or refocus their direction of thought. If as the producer you can tell that the session is going nowhere fast, then stepping in and giving your two cents might be exactly what is needed in order to kickstart the proceedings. Remember that the ideas and suggestions you make should come from a point of objectivity. If done with integrity and diplomacy you may well then be able to take a step back and allow the artist to explore and discover the best way forward, thus still giving them room to be creative while ensuring that the right focus is maintained.

Nurturing

Work alongside the artist, giving the time, space, and safe environment in which to try out their ideas. At first glance these elements might not seem like a big task, but nurturing someone's creativity is a difficult task to undertake as it can involve all the

(Continued)

other mindsets and approaches in one. This approach requires the skill of being able to carefully balance the encouraging comments and observations with the discouraging, knowing when these are appropriate and when they are not. This approach will also require you to have developed a sense of trust between yourself and the artist or group. They need to feel able to allow you to have creative input whether it be musical or otherwise. They need to see this help as a guiding hand and not a manipulative fist! A nurturing mindset and role could be considered as more relevant when working with younger or less experienced artists. However, all artists, experienced or not, can feel insecure at times and may struggle with their creative flow in the studio. It is at this point that they may look to you for guidance or advice and the objective viewpoint that is a fundamental part of your reason for being there.

Psychologist

This is possibly more an approach than a mindset and *psychologist* is by no means the most accurate of titles but more of a best-fit description. By psychologist we mean an approach that is more intentionally sensitive or switched on to the dynamics of a session or the personalities that are present in order to achieve a successful result. In many ways it is difficult to define this approach and single it out as a separate way of working, as it is used to a certain extent within all the other mindsets. However, this mindset or approach is worth considering when working in a potentially difficult situation or with an artist that is potentially more emotionally volatile. The ability to be perceptive and read and judge people's moods and body language can be a great asset for the producer, who can then react in the appropriate way in order to steer the session toward a positive outcome. Essentially what is being proposed here is that you adopt a more sensitive approach, allowing yourself to be open to these subtleties and consider how to react in a given situation. In doing so you may be better prepared to offer solutions to the artist and therefore create a better atmosphere in which creativity can thrive.

The role of the producer can require many different approaches or mindsets and is often an amalgamation of several blended together.

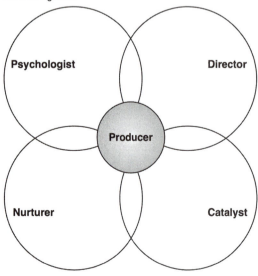

All of the approaches above and discussion as to creating a creative environment are generic to all producer roles or types and apply equally, no matter the skill set, knowledge, and expertise. However, if as an individual you possess the practical and theoretical musical skills, then your creative input during the session can potentially be quite considerable. Your musical ability to play, compose, and arrange will allow you to be a more tangible part of the creative musical process. This may be as basic as suggesting a slight alteration to the lyrics or the melody, through to playing or arranging and introducing new parts or tracks to the song to further enhance the production. The fine details of all of this will depend on the nature of the artist or band you may be working with and the way in which you have agreed to work together. As ever, knowing when and when not to use these skills is important, as some artists or groups may be more independent in these areas than others.

Creative and technical

As we know, different individuals bring a variety of different skills to the role of producer and certainly it is fair to say that some are more technical than others. By this we mean that they have possibly come from an engineering based background and are therefore at home thinking in this way and dealing with technical or sonic issues and requirements. So, generally speaking, the role of being both creative *and* technical within a session is only considered by those who fall into this category. The highly respected Hugh Padgham is a good example to use here. Padgham's production credentials speak for themselves but it is interesting to see just how much work he has engineered and mixed over the years.

Having the skills and knowledge on the technical side of the fence can give a very different slant to the session. This is not to say that a producer with a technical background will be solely interested in which kind of mic is being used and what the compressor is set at, but simply that they will be more aware and involved in the conversations and decisions surrounding the technical requirements. A recording engineer and assistant will be responsible for engineering the session and therefore much of the hands-on work will be undertaken by these people. Naturally there will be a degree of overlap with a producer who possesses skills of a technical nature; however, their main concern and focus will still be on the musical performance aspects. Therefore we're not necessarily talking mic choice here, but certainly mic positions for a certain sound or capture.

Do not be too quick to think that technical does not mean creative. As we all know, a compressor can be used purely to control dynamics or it can be used in a creative way to manipulate the tonal characteristics and emphasize transients or previously unheard nuances. A purely musical producer may know how he wants the vocal melody to sound but would leave the recording or mix engineer to interpret his requirements, whereas a more technical producer would not only know this but also be able to make suggestions or give directions as to how the desired sound should be achieved. In many ways the ability to be involved on the technical side of a production could be seen as an extra string to the bow. However, many producers would argue that they like to leave this to the engineers while they maintain a constant focus on the more musical and other aspects. Either way, having the ability to simply converse in technical terms can be a great help when working with your recording engineer in the middle of what might be a very creatively spontaneous session.

Sounding board?

As discussed in other texts regarding music production, a producer's role within a session does not *always* demand a great deal of creative (musical) input. Much of this will depend on the artist or band you are working with, their level of musical and instrumental ability, and the quality or completeness of the songs that have been written. This does not necessarily mean that you don't need to be a creative person in one way, shape or form, but more that you do not necessarily have to possess top drawer musical or instrumental skills in order to run the session and successfully produce the record.

Many of the producers we have spoken to certainly advocate that musical and instrumental ability can be an advantage; however, it is not always essential. Therefore, whether you possess a high level of practical musical skills and knowledge or not is not always an issue. So the question of what does the producer's role become if these attributes are not required needs to be explored.

In this case the producer's role within the session may take on a more *sounding board* type of role. By sounding board, we mean being the person who listens to ideas and thoughts and is then able to feed back with an objective opinion, offering suggestions, advice, and any other comments they feel might be beneficial to the process. Of course, most producers are sounding boards at some point in the project, as the very nature of the position requires taking onboard ideas and giving opinions. However, for a producer that may not possess a great deal of practical musical skill or is working with a musically skilled artist or band, their ability to listen, digest, give opinions, and direct (i.e., be a sounding board), will be their greatest strength and asset.

Within the session the sounding board role can be vitally important in giving confidence to the artist or performer. At times a quick "yes" or "no" may be all that's needed; at others a longer discussion to gauge an opinion might do the trick. If we think of the world outside of music production for a moment, this role can possibly be best mapped across to that of a trusted friend. We seek their advice not to necessarily be told exactly what to do, but more to enable and assist us in making up our own minds on a given issue.

While interviewing Rupert Hine on his experiences and work within the industry, his comments referred back to a time where he found himself being asked to produce an album without really understanding why. Rupert recalled that he was asked to produce an album by someone who had seen his name as producer on one of his own self-produced records. When asking what it was they wanted him to do they replied, "Do the same as you did for your own album; the only difference will be that I am singing on it." Rupert concluded by saying, "So I was never there to do a specific job, I was there to be myself."

It is intriguing to hear similar kinds of comments from other individuals that we have interviewed, and it doesn't take long to find a correlation in the comments contained within other books, where a guiding hand and a shoulder to lean on are cited as the producer's main input into a session.

It seems that, despite all the other skills and expertise a producer may have, sometimes all that is required by the artist is a like-minded person who they can bounce their ideas off.

BEING CREATIVE AGAINST ADVERSITY

It's all very well offering ideas of how to be creative and how to create creativity, but what happens when things start to go wrong? Wrong in a creative sense, wrong in a technical sense, and wrong in a personnel sense. Artists and musicians can be sensitive and temperamental souls and the close confines of the studio environment coupled with creative energy and egos can tend to magnify issues and heighten any underlying tensions that may be lurking. As the producer, you need to be aware and prepared to combat these situations in order to keep the session on track.

It is worth bearing in mind that, although we have chosen to split the possible areas of adversity into three main types (see below), there are many issues that may fall outside of our collectives. Indeed, some issues may seem to fall into one category when in fact the fundamental cause can be traced to another (for example, creative or musical indifferences that are actually caused by personnel issues within the band). Many difficulties that you may come across in the studio session are interrelated and are therefore not easily separated or straightforward. The following is intended as a guide and not a definitive answer.

Creative and musical issues

If things are going wrong in a creative or musical sense, the first question to ask is, why is this happening? (This may sound like we're stating the obvious but unlike technical problems that are generally easier to diagnose, creative and musical issues can be a little trickier.) If you *can* answer this question then you are halfway to solving the issue, whatever it may be.

Taking from our own experiences and those of the producers we have interviewed, many of the musical issues that occur during sessions seem to stem from either a misunderstanding of, or a lack of, the vision or direction for the session or even the project as a whole. Although you may presume that the prevention of this is to clearly discuss creative and musical intentions with the artist or band at the outset, do not assume that this will be enough to keep musical ideas on the straight and narrow within the session itself. As the producer you need to keep a constant eye on things as the session progresses.

A driver should keep both his hands on the wheel at all times. Taking one hand off or letting go entirely may cause the vehicle to swerve to the left or the right. Similarly, if a producer lets go or takes one hand off the wheel, the session may start to swerve in the wrong direction! This is not to say that the producer needs to rule the session and artist or band with an iron fist (as this approach would develop another set of issues) but that she should always have a guiding influence on the proceedings and remain in control. This is a little like when your Dad let you sit on his lap to drive the car… you had your hands on the wheel but he had his feet on the pedals!

So, as the producer you should hopefully be aware of the musical ideas that were being batted around and tried out and therefore you may have been able to see the issue developing. However, it is not always this easy and you may have thought that going down a certain road was a good route to take, only to find that the result at the end was not as successful as it could have been. At this point, being able to take stock and refocus on what is required is important. As unpopular as you might be with the band or artist, it is your responsibility to put your foot on the brake and consider where to go next. It may be as simple as taking a short break and explaining why you feel the prechorus is just not happening. Or all sitting in the control room with a coffee (or something stronger) to listen through the last few takes in order to gain a consensus of opinion. Whatever the musical issue, be it a guitar part, vocal melody, or lyrical idea, it is your responsibility as the producer to get to the bottom of what is (or what isn't) happening and put it right. It might not be your idea or solution that eventually puts things back on track, but this doesn't matter, your focus should be on what is best for the song.

As we stated at the beginning of this subsection, what appears to be a creative or musical problem may sometimes turn out to be a personnel issue, in which case a different approach may be required. We will address this within our personnel category.

One other main musical or creative adversity that a producer may have to deal with in the studio session is a lack of technical or musical proficiency on an instrument or from a vocalist. This is usually a sensitive issue due to the fact that ideally most producers would want the artist or group they are working with to be of the highest possible caliber. However, sadly this isn't always the case and consequently as the producer you may need to employ considerable amounts of patience and diplomacy when dealing with such issues.

The common sticking points are usually (in no particular order): the drummer cannot play in time and struggles with a click, the singer is out of tune or time, or the bassist or guitarist is struggling to nail the guitar part (or solo). The trick here really is to establish whether another factor is influencing or affecting their performance, or whether they are simply not able to deliver the necessary part. In terms of anything being out of tune, if it is an instrument and not a vocal then making sure that everything is in tune or tuned before the take is surely common sense. You might think it would kill the vibe to keep having to suggest this, but even a great performance is no good if it's not in tune. Therefore, getting the band or artist into a routine of always tuning up is worthwhile.

We have already discussed the importance of the headphone mix and how this can affect the performer in the previous section, so we won't cover this in too much detail here. However, suffice it to say if this is the issue it can be easily resolved by making sure that they can hear what they need to. Once this is done, the performances should improve, particularly with vocalists. If this is not the case, then things become a little more tricky both in terms of the practical solution and how you deal with the performer as an individual (as they are probably getting a little uptight after a number of failed attempts or takes).

Suggesting a break to talk through the part is a good option, although if the takes are numerous then a breath of fresh air and a change of scene might well help the performer clear their heads. (Again, much of this is all about communication and diplomacy.) If the vocalist is still flat or sharp in places, then getting enough takes to create a comped vocal part may be necessary. The success here could come from capturing the vocal in sections and a good engineer who is used to the comping process.

The alternative to this is to use pitch correction plugins; however, the amount that is needed might be considerable and therefore hard to conceal sonically speaking, making this a less than ideal fix. As a last resort you may need to consider getting another musician or vocalist to perform the part. However, this can cause serious problems for the original performer, especially if part of a group. This is where a group discussion is necessary to establish a way forward. Certainly replacing a lead vocalist is hardly ever an option due to the obvious tonal/performance differences that would be clearly evident. This makes the amount of time and patience a producer may need to expend on vocal tracking understandable.

Drummers and timing should go hand in hand, as it is what the job is all about. If a session drummer is being used then obviously this will never be a problem. However, if working with a band, getting the drums tight and locking into the groove may at times prove difficult. A drummer may have a great musical feel but their timing might not be as tight as expected or needed.

There are various remedies for dealing with this issue and, as ever, making sure the headphone mix is right should be the first port of call. Maybe more bass in the cans really helps the drummer to "lock in" for example. Tracking the rhythm section together "live" (i.e., drums, bass, and rhythm guitar) might also prove more successful in capturing a tight drum groove than overdubbing to a guide track, as this is the way that the band are used to playing their individual parts.

Further options that some producers may consider are playing to a click or a tempo locked loop. If the former is something that the drummer in question has no experience of, then it's probably not going to help a great deal. Listening to a very sterile click in the cans and trying to play a part musically takes some getting used to. Playing to a tempo-matched loop of some kind may prove more successful. If the loop itself has a good groove, then the drummer may be able to play along in time and play or feel the part in a more musical way, and therefore the performance aspect can be captured also.

Of course, as anyone interested in music production will know, there are now very accurate and reasonably quick ways in which a drum track can be edited and quantized, therefore making it very possible for an out-of-time drum part to be manipulated into time (Beat Detective, Elastic Audio, etc.). However, the result can sometimes lack real human feel (despite the ability to now dial this into a quantized part). The alternative to this is to comp a drum part together from sections of other takes where it is in time, hopefully achieving one complete 'in time' take.

With any of these issues a positive vibe and approach is often needed to raise the energy levels of a session, and here knowing what makes the people you are working with tick helps. (As mentioned previously, this is where taking the time to get to know individuals can really pay off.) Learning what makes the artist react can be a very useful piece of knowledge to acquire and can enable you as the producer to deal with these issues on a case-by-case basis.

With all these things, the need to maintain a perspective based on the initial intentions of the session is important. As the producer you are the objective keeper of the vision for the project, therefore while it's fine to allow the artist/band to explore different or new musical ideas or do just one more take, you are the person responsible for steering the path and keeping proceedings on track. Any deviation from this is your call.

Technical issues

Technical issues can be extremely frustrating to all involved, but especially to the artist or band who really don't know or care about what it is that has gone wrong and why. Their chief concern is getting a performance that they are happy with recorded. Therefore, one of the biggest issues with equipment failure, malfunction, or human error is the amount of time that is wasted hanging around waiting for the issue to be resolved.

From the producer's point of view (unless you are of the technical engineer type) you will most likely trust the technical complexities of the session to the studio's technical team and most specifically your recording engineer. If we think back to the planning stages, your technical requirements and resource and equipment issues would have been discussed and therefore any technical problems that occur within the session will be unforeseeable issues. The biggest point to make here is with regards to contingency. If something does break down, malfunction, or a technical error is made, then how can the issue be resolved quickly and efficiently with the minimal amount of fuss? If you are fortunate enough to be working in a professional studio, then the equipment being used should be well-maintained with a studio technician available should anything happen to go wrong. The chances are they can resolve the issue alongside the recording engineer and things should move along fairly quickly. If it is your own studio or home setup you are working in, then you and your recording engineer need to be well-prepared and have things in place so that you can continue recording should certain things happen. Specifics will depend on your individual setup, however, and generics like checking cables and connections, inputs, and outputs should always be done prior to commencing any recording session.

In this day of DAW software and digital recording technology, backing up is possibly one of the most important technical issues that should be considered and not overlooked. In pro studios, backups of all audio may be being made during the session in order that material is secure and will not be lost. If you are working with a smaller setup or home rig, then you need to make sure that either you or your recording engineer decides on a method of backup and how and when this will take place. To not have any form of backup is extremely inadvisable and to be honest there is no excuse, especially given that terabyte capacity hard drives can now be purchased relatively cheaply.

Lastly, if you're not particularly technical and come from a mostly musical background, then file management is probably not your favorite topic of discussion. It probably isn't even if you are technically minded! The file management within the session will most likely be taken care of by the recording engineer (or Pro Tools op). However, as the producer, understanding the way in which takes are managed, labeled, and transferred should be of some interest. You are running the session, so being able to keep track of everything can be an advantage and means that you are not constantly relying on information being provided by the recording engineer.

Personnel issues

Personnel or people issues are possibly the most difficult to work with in the studio. In fact, they have a tendency to affect many other aspects of the session and infiltrate the project as a whole in a negative way. The greatest skill a producer can have in these circumstances is that of diplomacy and perception. Being able to sense when there is an underlying issue and nip it in the bud quickly means that you might be able to avoid what could become some very difficult and unproductive sessions.

So what can you do about being able to improve in this area, as it is something that is very difficult to learn? Some would say you either have perceptiveness or you don't. We agree with this to a certain extent, but there are practical things you can do to help yourself in this area.

We have said at various points in the earlier sections of this book that getting to know the artist or band you are going to be working with is a wise investment of your time. This really is true when it comes down to working against some of the personnel difficulties that might occur when the red light is on. The session really is not the time or place to be settling disagreements, listening to arguments, or even counseling individuals with regards to issues in their personal lives. Nothing kills the creative process more, not to mention the vibe that you may have worked hard to establish. Therefore, try to find out a few things about the individuals in the band; we don't mean become their best friend all of a sudden, but simply try to gauge where they're at as a person and what kind of personality they have. It shouldn't take too long to work out who is the extrovert and who likes to take a back seat. This should hopefully prove to be useful when the recording sessions start. There's a chance that the disagreement over a musical part might not be wholly down to just the music side of things, but actually be because two band members are very stubborn individuals and both want things their way. If you recognize these traits beforehand, then you will be better prepared for these instances should they arise.

Knowing how to react to them and when to push a point and when to let go is important. Your goal is to get the best creative result you can out of the situation; how you go about doing this exactly is down to your own judgment and intuition.

THE PROCESS OF THE SESSION

In the last section we spent some time discussing pre-production and its importance to the session. How does this preparation come to fruition? How does it manifest itself? And what lessons and actions or habits should be carried on throughout the sessions?

In this section we look at the possible ways in which a producer can manage and run the ship during the sessions. We must suggest a caveat inasmuch as every producer will be different, as will every session, and every band you work with. So take the following paragraphs as a discussion of what you might need to recall later and how you might go about retrieving them.

Producer preparation

Preparing for the session is something we cover in detail in both project management and pre-production, but there will be personal preparatory routines you might wish to go through. The session can be a physically and mentally demanding sequence of events. The producer is expected to be serenely on top of matters and consider solutions, answers, and creativity when there's a block. You might have your own set of activities you like to consider when you're preparing for a session.

Some producers we have spoken to like to get into the studio and the session at least an hour before any musicians. Some like to use this time to sit quietly in the space, consider what went on the night before, and how they might work through any issues, or simply plan how the day ahead goes. The technically minded producers among them suggested they liked to use this time to review the recordings from the previous session and perhaps do some editing or planning the overdubs or whatever today's session has in store.

Turning up late can be an option for the more maverick producer. The band and engineer can get through the social pleasantries before the producer comes in and cracks the whip for the day at hand. This grand entrance can show who is in charge and will perhaps illustrate a clear signal that the session has begun.

Production team meetings

Some producers like to have a short coffee with the engineer and assistants before the act comes in the studio, just to set the scene and prepare for the day. The less technically minded producers would then use this opportunity for the engineering team to play back material and discuss a to-do list of what is ahead for them.

However you organize your times and work with your engineering team will come from experience or will come through necessity, as each project is very different. Formulating some quality time with your engineering team may help the more difficult projects flow forward.

Artist/band and producer meetings

By the same token, there may need to be clear briefing sessions over another cuppa with the artist so that they know what the day ahead holds, what they may need to prepare or rehearse for, or even write if they've got lyrics to complete.

Meetings such as these might be wind-down events after the session in the local restaurant or noisy bar in the city. However your artists wish to engage in the meetings might become the *modus operandi* for the remainder of the project. Alternatively, you may decide to lay down the law and insist on meeting the band in the local coffee shop, out of the studio, prior to the session each day to chat.

MANAGING THE PROCESS
Team dynamics and your interpersonal skills

As we introduced in Chapter B-2, Your People, the team is paramount to the success of any production. The team surrounding the project should have a dynamic that will drive the music forward and the production as a whole. Understanding team dynamics and the way in which the project can be best realized within those parameters is an important facet. Producers will choose to work with one of a few engineers they've enjoyed working with before. Yes, these engineers need to be of a professional standard, but if an engineer is a pleasure to work with, gets the job done and also gives a great vibe all round, it is more likely that he'll get the phone call next time around.

Tim Speight comments on the way in which he has been contacted for work and highlights that your interpersonal and communication skills really do count. "You can get gigs on the strength of what you have done before, but what I've found is that I have probably gotten more get gigs from (a) doing the job, but (b) (without wanting to blow my own trumpet here) personality. It's about personalities within the studio. I like to get into it, I enjoy it and have a laugh, and I can communicate with people. The best sessions I have ever had is where you're not really aware that there's a deadline and you're not aware of the "product." Everything just works and everybody just gets on with it, the vibe is right, and it just comes together. I have gotten a lot of work from people who say that they prefer to work with me as the job gets done but at the same time it's enjoyable."

Rob Orton describes some of the interpersonal skills needed in the studio environment. "Communication skills are such an important thing and people skills really. In a studio it's such a strange environment that one of the key skills is to be able to have the foresight into what is going to be happening next; you get that from clues. Quite often when you're in the studio you figure out what's going to happen next by overhearing a conversation, 'somebody said something so that means in a moment they're probably going to want to do this'. There's a lot of deduction. One of the really key things is perception. Sometimes somebody tells you something they think about a mix. The thing that they say doesn't make sense but you've picked up on a little thing about the way they've said it and it gives you the clue about what they want. It's a really key skill I think for pretty much every role in the studio."

Most producers will be masters at ascertaining the skills and personal dynamics between the band. They will have many tricks up their sleeves to be able to work with a variety of personalities in a range of situations. They will consider the process, and be continually sensitive to the needs and feelings of the band. It is a joint venture after all.

Being considerate is an attribute most of us respect in others. In our work and through the research for this book, we have had the opportunity to meet with many producers who have had fantastic demeanors. They are some of the nicest people you'd like to spend time with in the studio. It takes a breed of person to be like this. It would be wrong to suggest that this is a common feature of all producers as many will work on different models.

Whatever your methods and way of working, don't underestimate the difference your approach and ability to deal with differing dynamics can make. You may be running the session but you certainly can't do it by yourself!

Getting the take

The session has started, you've listened to the previously tracked parts a few times through, the vocalist knows the song, and all you need to do is get the vocal tracked and it's in the can. But how do you get "the take"?

This question can be asked of any instrument or part but for some reason the hardest track to be satisfied with is often the vocal. This is possibly because the human voice is such a natural and expressive instrument that any inaccuracy or falseness will be identified more readily even by the least discerning of listeners. If you're fortunate enough to end up working with an artist or band that has great vocal abilities and has recorded before, then this process might not be as difficult as we are portraying here. However, if the vocalist doesn't have the best voice you've ever heard or hasn't recorded a great deal before then, how do you go about capturing the best take?

If you think about what really affects a musician's performance you will probably be able to come up with at least four or five different influential factors. For example, the song's meaning, the recording space or environment, their physical and mental state, and the people they are working with. We've covered many of these issues in Chapter C-3, The Desired Outcome, but what do you do when you're actually in session and the red light is on?

Many engineers and producers we have spoken to have mentioned the word *vibe*. The problem with getting the vibe right is that it really varies depending on so many different factors. However, there is a very commonsense way to approach this and at least get the basics right. Put yourself in their shoes; how would you feel standing in that studio's live space or vocal booth singing that song? What would be the things that would bother or affect the way in which you perform? Asking these questions should bring forward a number of things for you to consider and be aware of. (This could all be part of your preparation before you go into the studio for the session.) As long as you are aware of potential issues, then that's a good start. The specifics will be down to the individual vocalist themselves. For example, one may want the lights up, the other may want them dimmed or virtually off! It's whatever works in order to get the best take possible; the key is to be perceptive, open minded and creative.

THE VOCAL TAKE AND LYRICS

When tracking the song and overdubbing the vocals, it can be hugely invaluable to have copies of the lyrics at hand. Ideally, photocopy or print off several copies. Consider using the highlighter technique when tracking vocals:

- Ask the artist to provide you with numerous (preferably typed) lyric sheets.
- Take three colors of highlighter.
- Use the first (lightest color) to denote any areas that have issues with them in the first take.
- Use the second lightest color on the same sheet to denote second-pass issues.
- Use the darkest color to do the same on the third pass.

Using this method, you can quickly see the hot spots needing attention. Keeping fast notes can relieve you to concentrate on drawing the best vocal performance possible from the singer. Simply cross out the lines as you comp that perfect take together.

Tim Speight explains how he dealt with getting the take in terms of vocal tracking from a new or relatively new artist.

"If an artist hadn't been involved in the songwriting process and a lot of the time within the manufactured stuff they might not be, what we would do at that point is get a session singer in as a guide. Someone who I knew could nail the emotion, the phrasing, everything to do with that song to start with… that's the first thing. When the person comes in and you're trying to get the real vocal take from the artist, this is where you've got to allocate quality time to take the pressure away from them. We used to get a lot of artists who were relatively new and they'd be petrified coming into a pro studio with all the gear, et cetera. It's sometimes got the potential to go wrong. What we used to do when the time was available was get them in but tell them that 'we're not going to record today, we're just going to have a mess about with the track'… and the relief on their face. We didn't record them we just got them into the control room. What you're doing is setting the scene and getting them to relax and have a laugh with it. If you can just break down that barrier, get them to relax, and get away from that phobia of going into the studio. As an engineer or producer you're going into the same studio every day but for a vocalist it is different. I have a lot of respect for vocalists as they are projecting their own instrument; they don't have anything to hide behind."

Liquid refreshment?

The folklore exists. Yes, the band is drinking a crate of lager throughout the late going session, believing that the album is sounding better and better. They might escalate the alcohol levels, or indeed go through to something altogether different. It happens less these days, but we'd be naïve to say that it does not happen.

Prepare yourself. You have some choices. You can either indulge in some or all of what's on offer (providing you can handle yourself). There are those that have suggested that it is imperative to be in the same "space" or "zone" as your musicians; to hear what they hear and to share the same responses. This is your choice.

You can take the view, of course, that you're above any form of indulgence and therefore in control of the session and the mother ship! In this state of mind, you're more likely to be able to objectively know when to accidentally-on-purpose undo that last Pro Tools take, as the performance drops. Conversely, you might find the artists inspired beyond their normal levels and instruct the engineer to simply keep the tape rolling.

Recording the recording

It is a common opinion among music production students that, as musicians, engineers, and producers now use DAWs such as Pro Tools and Logic, the need to take notes in a session, use track sheets, or make production logs is obsolete. In some ways we can see the argument for this, as DAW software has the ability to store and recall so much information. However, what is being forgotten here is the need to have an organized approach to what you are doing without having to rely purely on the instant recall of a Pro Tools or Logic session. What was recorded on which track? Which take was best? Which mic did we use and where did we put it? What reverb was applied? These are all details that could be noted down on a track sheet during a session. The ability to recall this information without having to be at the computer running the DAW software session could be really useful at some point in the production process, especially if you're discussing the session and its detail outside of the studio, for example.

We are not suggesting that every professional engineer and producer make a huge amount of written notes in every session, but more that they will not just rely on the recording software to record and catalog every choice and decision that is made. How and to what extent you decide to do this is a personal choice. It's as much about developing the right mindset and approach to a session as it is about making a note of all the details.

In days gone by, mixing was something that usually the recording engineer did. It was nothing too fancy, just a little bit of creativity through the careful balancing and blending of precious and carefully recorded material. However, the art of mixing has changed as individuals within the industry now class themselves as purely mix engineers, and mix engineering has arguably become a definitive art form in its own right. These individuals include Chris Lord Alge, Bob Clearmountain, and Rob Orton, among many others.

Choosing whether to mix within your immediate production team (whether you mix, your engineer, or a combination of the two), or select one of the leading mix engineers will be a decision to make along the way, or indeed the label or artist may make the decision for you. Nevertheless, the mix can be the most exciting aspect of the process for some engineers as it is the final shaping of the music, the image that will live forever and formulate the product attached to its genre and time. In this chapter we dissect the mix and how this can be produced and approached.

WHAT'S ALL THE FUSS? WE'LL FIX IT IN THE MIX!

The mix can be considered as the product. In some ways it is quite perverse that so much value is given to the mix, but this aspect, alongside mastering, is the longest lasting image of quality which originally started with someone's performance. This image can last generations, represent movements, and generate emotions and time stamps to many listeners.

The mix (alongside its sister process, mastering) might be likened to the paint job on a car; the motor car is still an impressive mechanical invention even by today's standards; however, it is the paint job, the styling, the badge, and the image that give its identity and assumed quality, whether that is replicated inside the mechanics or not.

Sticking with the car paint analogy, it could be misconstrued that a car, however bad, that has a great paint job, will outstrip the competition in terms of performance. This is not going to happen. The same is true of recording and mixing. We've all heard the expression, "We'll fix it in the mix." This has become, to some, a real impression (some might say epidemic) of what can happen these days.

What is Music Production. DOI: 10.1016/B978-0-240-81126-0.00011-1

Mass computer access and high-quality systems such as Apple Logic and Pro Tools LE have opened up so many possibilities. Many artists and producers have retreated from the *modus operandi* of hiring a large, needless to say, expensive recording studio when their front room or garage could suffice. However, so many of these recordings can fall short of the sound quality expected. This is where a solid mix engineer can make all the difference and, yes, they may on occasion have to "fix it in the mix."

However, this cannot be simply attributed to recording in a nice studio. Perhaps it is the skill and ingenuity of the traditional production team that can be lacking in home-based projects. It is now much easier to develop skills to circumnavigate a good acoustic space and good skill at performance, but this does not always live up to the real thing. As previously mentioned, we believe that the CAP (capture, arrangement, and performance) is a good way of achieving high production levels. As such, a well-recorded and executed production in the recording sessions will probably ensure the mix is of a higher standard whatever happens to it later down the production line.

AUTO TUNE, BEAT DETECTIVE, ELASTIC AUDIO

A brief word here about the use of certain audio manipulation processes. It is very true to say that "fix it" software has come a long way in the last few years. We only have to consider the eye-watering capabilities of the latest version of Melodyne to realize that we really can fix a performance whether it be a monophonic vocal line or a polyphonic banjo part! At one time options within such software were limited and the resulting sound after processing was far less than transparent, especially if the process had not been implemented that subtly or accurately. Now if handled well and used sparingly, it is increasingly difficult for the average listener to detect. (Of course, this is not to be confused with the deliberate over use of such processing where the intention is to be able to hear the creative sonic effect that is caused.)

However, it is important to realize that even if this software does allow us to fix issues in a reasonably transparent way, this will still not alter or enhance a performance that is lacking in either musical skill or performance energy. Plugins for these aspects unfortunately do not exist and probably never will, so the need to capture a high-quality performance at the time of recording will never lose its gravity. Any good mix engineer will certainly be able to make some impressive fixes with the tools and expertise that they have at their disposal. However, those that we have spoken to in the process of writing this book have all stressed how much of a difference getting the right take in the first place can make.

Grammy award-winning mix engineer Rob Orton comments on the issues of fixing it in the mix: "More can be done in the mix but I think it's a terribly bad idea to try and do it all in the mix. A mix generally takes a day and you're looking for all those issues and key decisions to have been ironed out. You're just trying to present what's there in the best possible light and sometimes that means you've got to dive in and fix things, but all the time you're doing that you're taking away from the time that you could be spending on other more creative processes."

WHAT IS MIXING?

This book is not intended to be a how-to guide, but, as the introduction says, a guide to guidance. Therefore, it would not be sensible for us to try to teach how to mix technically speaking, but to look at the production strategies one might employ to mix with.

If you really want to get into the process of mixing in depth, there are many good books, such as Izhaki, R. *Mixing Audio: Concepts, Practices & Tools* (Focal Press, 2009) which offer a wealth of insight into the world of mixing. The same is true later with mastering. A good reference here is Katz, B. *Mastering Audio: The Art and The Science* (Focal Press, 2008).

If you consider for one moment what mixing entails, we can suddenly think very differently about the process. In raw nuts-and-bolts terms the process is simple: mixing takes 24 channels (sometimes less or more) of different levels and treatment through some kind of summing system down to 2-track stereo (or surround if required). This is a simple process and something we probably take for granted. However, just think of the millions of permutations available simply using the faders.

On a technical level, mixing can be thought of differently. Its art form, as described above, is to *squeeze* the 24 channels of recording into a 2-track stereo mix: 24 fully dynamic recordings of parts to be managed, blended, and polished onto two tracks. It is a considerable expectation for this to be easily achieved, and it is simply amazing what is possible when a track becomes something larger than the sum of its parts.

However, consider this differently yet again. Through mixing, the production team have the opportunity to take a snapshot of a time and place and deliver this onto a canvas, a little like a painting if you like. A scenic painting encapsulates an image of a landscape at a particular time and place. Due to the scale the artist employs, it offers a representation of something much larger than the physical dimensions of the canvas, evoking different emotions captured in a different age (think Constable or similar). Mixing is an art, and can be considered in this way for it captures the time (the '60s, '70s, '80s, and so on), an emotion (whether personal or otherwise) from the age and genre.

Mixing is therefore important. However, if each producer was to continually consider the cultural impact the mixes might have, they'd perhaps fail to make so many decisions in quite the same way which make our music scene so dynamic. Spontaneity is still as much an important ingredient to the creative process as when recording the tracks.

However, a vision for the mix, as with the whole production, should perhaps be considered. In Section C (see chapter C-2, Pre-Production) we discussed the

V.I.S.I.O.N. for a production; mixing still should follow this vision. Keeping an eye on the desired outcome and working to this might ensure a successful product, or should a producer not realize new scope for creativity when it presents itself, might restrict the overall possibilities.

Visiting the technical level again, mixing is, of course, much more than simply throwing up the faders in an organized arrangement. Much development and best practice has developed over the past 50 years. Key components remain, such as equalization, gating, compression, limiting, reverb, delay, and, of course, level and pan.

As with every art form and professional process, development has taken place in this arena also. We can always think of revolutionary developments that have improved our ability to mix, namely automation, audio editing (DAWs), and tuning tools (as previously mentioned) being just some. Alongside the technological-based developments are also the practice-based ones.

Each decade has seen not only its own style of sound shaped often by the equipment around it, but also practices in the studio have been set and broken time and again to create something new. Just think of the 1980s, for example. The bass was often not as big as we now would like, but the recordings were often precise and quite bright. Why? Well, we are not all that sure, to be honest, but improvements in equipment design, noise reduction systems for analog tape, and the use of more synthesizers, both analog and digital, may have improved the treble end. Culturally new musical movements, certainly here in the U.K., meant that the music was stripped down, leaving more space for clean and new digital reverberations to be heard. Each decade has its sound, both musically and in terms of engineering. Perhaps only now do we have a scenario where the technology may not define the sound of a decade.

Mixing should be something that flows naturally from the intended arrangement as set out by the producer and artist. However, these days mix engineers often edit a considerable amount to find the best vocal take, or replace or layer drum parts with better sounding samples. This aspect of mixing has become quite the normal practice in professional circles, certainly for pop recording. Every aspect of the recorded material is assessed for its perfection and repaired accordingly where necessary. Vocal tracks, both lead and backing, might be compiled from many takes and then tuned manually in Anteres Autotune (or similar). New parts might also be tracked to add effects and fill a poignant space in a recording. The art of production is still very much part of this mix process too, and can often be without a great deal of input from the producer.

VISION FOR MIXING

Mixing can be hard at the best of times, much more so before the option of total recall. When mixing on a large format console, tape machine, and so on, producers and their engineers could not afford to have commitment issues. Studio time has always been expensive and therefore has needed to be results driven.

By the end of the booked time in the studio complex, the production team would have liked to have delivered the mix for the singles and album. For some, perhaps running out of budget, there was no going back. To set up a large format console with 20 or so tracks of audio and a MIDI sequence tagging along too with any number of synths feeding in for a mix is quite a large undertaking. Without total recall, coming back to the same point can take a matter of hours, and hence is an expensive option.

In those days, it would be common for production and engineering teams to have to commit to tape. We've spoken about this before, in an age where tape ops were actually given an incredible level of responsibility. If they pressed "record" too soon they might wipe over the good lead vocal part from the previous section before dropping in for the new chorus. There are, of course, much scarier stories in circulation. The production team became confident professionals working within the bounds their equipment would allow them.

Producers would work on details from the start, ensuring positive performances and that each take counted. Producers were perhaps more willing to make confident decisions, as there were not many options for nonlinear editing other than splicing the tape, and across 24 tracks this is not always easy. These confident decisions were often borne out of an overall vision for the mix. Again this comes back to the vision the producer and artist have for the album as a whole. The producer should retain the overarching strategy for the production, taking into account the ideas provided by each stakeholder.

It is fair to say that the technological changes that have occurred within the music production world in the last five to 10 years have altered not only the way in which producers record, but also how, when, and if decisions are made. The ability to add more and more tracks and keep various takes until the very end means that decision-making is possibly not as pressured as it once was. This in turn has affected the size of the multitracks (in terms of track count) that some mix engineers have to deal with from time to time. This makes their ability to be decisive and discern a vision for the mix ever more important. If, as the producer, you have had, and stuck to, a clear vision for the track then you will have most likely made many of the necessary decisions along the way, thus making the mix engineer's life a little easier.

The innate strategies for mixing can be considered on two levels: The strategy of the producer as we've discussed above, looking at every aspect of the production process, the mix's place in the genre, market, and so on; and the strategy of the discrete and individual mix. Much of the strategy for each mix will be based on the engineer involved. Producers may choose to use their own engineers based on outcomes and relationships, but today, as previously discussed, there is a small army of emerging specialist mix engineers available to work remotely on the track.

When working with a specialist mix engineer the producer will often send a reference mix, possibly put together by the recording engineer, in order to give a flavor or direction as to where they hear the mix going. To the mix engineer this can be used as a rough template or guide, a road map for certain ideas and possible mix "direction." It is important for the producer to communicate her vision of the mix to the engineer, as this will help further guide the process. What elements need to be picked out and focused on? Where are the hooks? How should they be enhanced and brought to the fore? How the mix engineer actually addresses these issues is down to their own judgment and interpretation (the methodologies behind how they do this is discussed a little later in this chapter), and it is this that can really add something different or special to a track.

Tim Speight's comments not only highlight the importance of the mix engineer and the changes that they can (or sometimes not) make, but also that of the reference mix and the mix engineer's interpretation of it:

> "In the hierarchy of everything, I always think of the mixing engineer as being on the very top, because that's where ultimately the raw material is pulled out and you've got to make it balanced and make sure that everything is all in there.

> "I have had it a few times where as the recording engineer I will do a 'board mix' for the mixing engineer to reference to and they've missed it... they've not captured something. At the time I was just vibing with it and just doing something and before you know it it's working. But you don't know what you've got sometimes until somebody else goes away with it and you think, 'Well actually where's all that bit that we put in there?', that's buried down in the mix now? He is hearing it completely differently. This is where you need the producers to come in at the end and rebalance it."

The mix engineer is always looking to improve on what has already been achieved, and although some changes may need to be made at the producer's, artists' or label's request, the goal is always to present the best possible result for commercial release. Therefore, if you, as the producer, want to make changes you should be able to discuss this with the mix engineer and end up with a result that not only pleases all those concerned but that also sounds great.

CHOICE OF ENGINEER(S)

The choice of mix engineer for a project can be based on a number of different factors and there are various stakeholders who need to be considered. Much of a mix engineer's work arrives via a record label or producer and therefore, although you as a producer may have a particular mix engineer in mind for a project, you may also have to contend with the wishes of the record label and/or the artist. It is not unheard of for record labels and artists to request the skills or name of certain mix engineers as these can carry a certain amount of kudos and may provide the hit that all concerned are looking for. A label and their A&R will look at and consider the style, genre, and market that the record is intended for. Therefore they (or you as the producer) may listen to other similar style or genre tracks, like the sound, and request that the same mix engineer be used on the current project. For mix engineers this means that their previous work (mixes) is often the measure by which they are judged and therefore hired. This is something that was certainly evident for both Tim Speight and Rob Orton when they spoke to us for this book.

As we have already discussed in Chapter B-2, Your People, much of the reasoning behind the choice of mix engineer for a project will be based on relationship. Many of the partnerships that have been formed between producer and mix engineer in the past have occurred due to a natural working partnership where trust and creative flow have developed. This is obviously very similar to choosing a recording engineer where trust, communication and creativity are so important during the sessions. The biggest issue here for a producer is trust. If we consider that a producer may have spent a considerable amount of their time crafting and fine-tuning the production of a track, the person he then finally hands the track to work on, needs to be in the same ballpark, as it were. There needs to be some common thinking between the producer and mix engineer, and their ability to communicate their views and ideas to one another is vitally important if time and money are not to be wasted.

The other deciding factor will be the specific style or experience that a certain mix engineer can provide to a project. Most successful mix engineers will be able to deal with a wide range of styles and genres, but some may specialize in particular areas. This may not be through their own intentions but more that they have achieved specific and successful results in the past, which labels and producers then lock onto. They therefore become sought after and to a certain extent may always tend to get asked to mix certain types of track or artist.

MAKING SOME DISTANCE FROM THE SESSION

If you are producing the record and have therefore been working alongside the artist(s) for some time, you will have become close to the project. In some ways this is why mix engineers have become so attractive—not only because they can help sell records but also because they are another fresh perspective on the project.

Mix engineers today bring the unique flavor of their own style of production to the existing project and the potential value should not be underestimated. In many ways the mix engineer can refine and enhance an existing identity or redefine an existing identity into something different. There are many good examples. One rather obscure one would be Nerina Pallot's hit "Everybody's Gone to War" which as a single is pretty good, but it was Chris Lord-Alge's mix that made this happen. Another good example of this is the commercial and radio-friendly sound brought to the Amy Winehouse track "Rehab" by mix engineer Tom Elmhirst. Here Elmhirst takes Ronson's retro late '60s soul sound and merges it with a much more contemporary hip-hop flavor. His use of kick and snare samples layered on the track assist in bringing the sound up to date, while the use of spring and plate reverbs complement the production style that Ronson has intended to capture.

If a producer is to effectively play a part in mixing the project or in directing the mix engineer, then the need for objectivity and distance from the song is essential. Many producers that do not take part in this process use the opportunity to do other activities while coming back to the project for critical listening, further direction, and approval. The amount of creative input that the mix engineer can provide can vary, but ultimately the responsibility of realizing the original vision lies with the producer.

Tim Speight prefers this way of working, saying, "As a mix engineer this is how I like to work, I don't like the producers to be there while I mix. I prefer them to come in fresh at the end once I've got the mix going on and make some final tweaks."

YOUR INFLUENCE ON THE MIX

The producer's influence on the outcome of the mix can be varying. There are stories from the studio that suggest some producers walk in and declare a number of changes, and when out of the room the engineer recalls the previous settings. Some producers will be happy with the attention that they have paid to the arrangement and the instrumentation (the CAP) and therefore feel able to let much of the intricate mix work go to a trusted engineer. However, there are many producers who wish to remain with the work, working the production team hard until the mix quality is reached.

As we have already mentioned, the label's A&R will most likely want to hear the mix and will certainly have an opinion about it. Therefore, it is important to remember that the process can involve a number of parties who will all have a vested interest in the end result. The main thing to consider here is the level of compromise or certainly discussion that may be needed. Again, clear communication and understanding will help, but as the producer is the project leader as it were, their views and opinions should be heard and trusted by the other parties involved.

Of course, there is absolutely no reason why the producer can't be entirely happy with mix coming from the mix engineer. After all, the very reason that they are involved is because of their abilities to produce great results. In this case your input on the mix may be very minimal!

MIXING METHODOLOGIES

Many engineers have their own mix strategies or methods, which could be as simple as "bass drum up" or "vocal down" mix. The "bass drum up" method is where many engineers begin, starting with the bass drum and then snare, working up the kit and bass guitar to build a solid background for the vocals to finally sit on. Other engineers work on the lead vocal first and then spend time building a collage around it, shaping and layering the sound to fill the remaining space. Neither method is right or wrong and much of the decision should be based on the individual track and what element is considered to be of primary importance.

A mix engineer employs a high level of listening skills in order to discern the various elements of a track and how they are being integrated within the recording and production. Tim Speight discusses the importance behind being able to critically listen to a track and assess what's going on in order to formulate ideas and an opinion. "One principle that I've drawn on is the idea of 'Christmas tree mixing' where you try and see the music rather than just listening to it. It's amazing when you start analyzing the layers of music from the foundation with drums and bass and how everything layers together and links and builds, along with the width and depth. All of a sudden you can really start to see it."

We are sure there are other engineers such as Michael Stavrou (author of *Mixing with Your Mind*, 2004) who often begins with the part that interested him the most. There are often elements within a track that are unique and could be considered the nonvocal hook, whether it be a drum pattern (Soul II Soul), a vocal, a guitar riff (such as in "Layla" by Eric Clapton), a bass line (Queen's "Under Pressure," also sampled by Vanilla Ice for "Ice Ice Baby"), or simply a unique texture made up of a number of elements. These elements can be more memorable than the lead vocal in some cases. Therefore, it can be sensible for this prized element to replace the vocal in a vocal-down mixing strategy. As we've said previously, the decision to approach the mix in any of these ways will be based on the vision that is established during the initial listening of the reference mix and material.

Rob Orton describes the approach to mixing from his perspective: "First of all I always listen to and pay a lot of attention to a reference mix and I think that's where you really start. By listening to that you get a sense of what's intended. When you start a mix you're looking for all the little hooks to draw out of the song, and you're figuring out a way that's going to draw people in. When you make those decisions you're trying to decide what's important about a track, quite often it's a vocal or it might be a slightly odd balance or something. It's a difficult one to describe as it's always something different. A big part of the job of a mixer is trying to find that 'thing.' It's about sitting down and working out what's good about it, what's not so good about it. How can you make the things that are good more prominent and how can you make the things that aren't so good less prominent? That's a big part of it."

As many producers come from the musician-producer ethic they are not necessarily technically minded. These producers may take the opportunity to rest their ears during the mixing of a track by allowing the engineer to work alone. The producer can then return and provide constructive critical feedback based on a fresh set of ears.

Engineer-producers will likely be in the engineering hot seat themselves or will be working closely with a chosen engineer on the mix. This may be the position you find yourself in, depending on your own skills and experience. In this case the communication will be immediate and a good working relationship is needed.

Each strategy for the mix based on the producer's involvement will be different. One thing will be assured: the producer will be aiming to hear a mix he is happy with firstly for himself, but also for his clients, whether they see that as the record label, artist, or both.

BEING HAPPY WITH THE MIX

Being happy with the mix should be something that comes relatively easily, given that as a producer and/or recording engineer you would have taken care during the tracking process to gain the best capture, performance, and so on (producers CAP), and then a mix engineer guided by your reference mix and input would have brought the best out of the track sonically speaking. If this is the case, then being happy with the mix is not really an issue and it's a job well done.

Where the mix begins to be an issue is either when original ideas don't sound as they were intended or someone changes their mind and requires something different from what was originally agreed. The important issue here is to establish why this is the case and what needs to happen to rectify the problem. If things do not sound how they were originally intended, then it needs to be established whether this is something that has happened due to changes made in the mix process or whether the mixing process has simply highlighted a weakness that already existed in the track.

Perhaps the most simple issue to resolve is one of the mix engineer not treating the track as was expected or intended. It may be that they have highlighted different elements within the track, therefore affecting the blend between parts of the arrangement/production, or have simply processed something in a way that does not quite sit right when compared to the producer's or artist's original vision of the track. Either way, a chat to the mix engineer going over the issues and the necessary changes should not be a huge problem, although obviously another mix will take more time. How much more will depend on the degree of change that needs to take place. If this is only a few minor tweaks, then it may not take much time to rectify at all. Most mix engineers will be happy to make tweaks and minor changes to the track once feedback has been received, and normally the time taken to do this would not be charged for. This is an aspect that would be discussed and finalized with the mix engineer when contracts are agreed.

If the final mix highlights weaknesses in the original recording that may have been missed or weren't as obtrusive before, there are two possible routes to take. (We would like to point out here that a good mix engineer would do their utmost to rectify any possible issues during the mix process; however, this may not always be possible.)

Firstly, there's the "fix it in the mix" approach. This is by no means a great way to proceed, for reasons that we have already covered and the end result might still be short of the mark, given that there is only so much a mix engineer can do. However, depending on the issue, this might be a perfectly viable option (one that the mix engineer might have already taken). The alternative route to this would be to rerecord whatever it is that is causing the problem within the track and hopefully iron out the weakness that has become apparent. This solution will not only take more time and incur extra costs, but will also require the artist and musicians to be available once more for recording. This is the reason why so many producers and recording engineers will always pay so much attention to detail and quality at the tracking stage.

If you fall into the engineer-producer bracket, then you may decide to mix the track yourself or possibly alongside your recording engineer. Although we have mentioned a lot with regards to the specialist mix engineer, it is still quite common for a recording engineer to end up completing the final mixes, possibly with or without the producer. (Tim Speight and Hugh Padgham are again good examples here.) This is obviously very much a personal choice and depends on background and expertise. However, it is worth pointing out that if this is the case, then you may find it harder to maintain a level of critical distance given your close involvement in the recording sessions and overall production process thus far. Having said this, you will most likely have been steering the recording process very closely with the final mix in mind and therefore much of the hard work may have been done already.

The important issue here is knowing when you are happy with the mix and that you have done all that can be done to present the track in the best possible way. Knowing when to stop tweaking and call it a day is important, as after a certain point you may not be making any further improvements at all. Worse still, you may actually be undoing what you have already achieved. One analogy that may be helpful is that of climbing a mountain. When starting at the bottom there is a long climb ahead, but as you climb you are constantly making progress. Eventually you'll reach the summit and be at the highest point. After this, the only way is back down, possibly even retracing your steps!

The mixing process can be very much like this at times, with the first few rough mixes making good progress until you reach, say, Mix 4 when after this the quality starts to tail off and you're actually heading back down again!

Most of the decisions with regard to the final mix should be made with the best possible result in mind. The route that is taken to achieve this is to a certain extent secondary as long as it serves the purpose.

MIX EVALUATION

There will be down time from the recording or mixing sessions, whether that be for a cup of coffee or a longer break. Breaks are excellent to rest your ears and rejuvenate you for a refocus at another time. On returning from a break, your ears are at their most objective and it is at this point that a listen back is most useful in determining issues and the next job on the list.

Developing a personalized system by which notes can be taken quickly without drawing too much focus away from listening is recommended. The following is an example, but you should use whatever system you develop.

Time	Symbol	Description
00:01	–	Poor start - timing
00:42			Guitar out of tune
00:59	O	Issue with cohesion between bass and drums
01:02	–	timing issue with double bass drum here
01:32			vocals again
01:39			vocals
01:47	O	issue with cohesion between bass and drums
02:03	O	Issue with cohesion between bass and drums
02:41	–	Middle 8 lacks punch and cohesion
03:20	–	Timing issue at end of piece.

It is good to have a personalized method by which notes about specific issues in a mix can speed up workflow. On the first pass, symbols can be used to note there is an issue, and descriptions can be added later on subsequent passes.

The symbols in this example make the notation quick and easy to take down on a first pass, especially if there is a lot to notate in a short space of time. The symbols here show a " – " for timing issues, " | " for tuning and " O " for any cohesion issues needing attention. Later passes can be used to embellish the descriptions if necessary.

If you are running the DAW, then it will also be possible, say with Pro Tools or Logic, to enter markers and simply place the " – " or " | " or " O " in and come back with further descriptions on a second listen.

ROB ORTON ON BEING A MIX ENGINEER

The following is from an interview conducted with mix engineer Rob Orton while researching for this book. Our intention behind including this excerpt is that it provides some professional and individual thoughts on the topic of mixing.

Commenting on his relationships with producers and where his work comes from as a mix engineer: "My best example of a producer/engineer relationship was probably with Trevor (Horn) when I worked for him all those years (as one of his recording and mix engineers). Now as a mix engineer I guess I'm building relationships with various different producers. But a lot of the work actually comes directly through from the record companies or label. Quite often it will be because the label is not happy with a mix that a producer has delivered."

Is that a mix that they might have delivered/done themselves or has that come from another mix engineer that they have been working with at the time? "It varies, sometimes you get stuff that has been done by the producer and I guess the producer is probably expecting it to be handed out to a mixer. But quite often it's the label that decides who's going to mix something, it depends on the producer. If you've got a really big-name producer then they have a lot more sway."

What do you think the things are that A&R departments or labels are looking for when they go to a mix engineer? "It really varies. Sometimes people have got a rough mix that they like the vibe of but they're looking for it to be just a bit better. Sometimes it's a question of taking it to the next level but not changing it too much, and other times you're looking to bring a new angle to it where it's not quite right and it just needs another take on it. In that way they're looking for another creative input and something a bit different from it, but to be honest that happens more and more rarely actually. I suppose because more and more people have got Pro Tools rigs and stuff at home so the standard of rough mixes or demos has gotten better in some ways."

How has this affected people bringing work to you? Does that mean that people are trying it more themselves these days? "There are two effects really, one is that you get this thing happening when people are really attached to a rough mix or demo, and because quite often a songwriter will be a producer as well these days they don't necessarily always have the skills to finish it off, but they have a creative kind of brain that takes it a considerable way along. So it ends up at a point where everyone is really enjoying the vibe of something. Then it comes to me and that makes my job a bit more difficult in some respects because it restricts my freedom to be creative. On the other hand, that means there's quite a lot of work for guys like me, because sometimes it needs quite a lot of sorting out. If it is done by someone who is primarily a songwriter/producer rather than an engineer there's quite a lot of technical issues with it that need fixing. A lot of the time that is what I am paid for really. This just means that it has placed a greater emphasis on the importance of a mix engineer."

CHAPTER D-3
The Mastering Session

"Mastering is the skill of modifying the sound of a mix to give the best reproduction possible via its delivery format (CD, vinyl, et cetera) and the service to produce the final master(s) from which all the manufactured copies will be made. It is also the last opportunity for any artistic input by those involved in the production, so the mastering engineer should be sympathetic to the sonic quality that the producer and artist wish to achieve. Processing may be required for a variety of reasons: Mixes may have been made in a poor acoustic environment or with substandard monitoring; optimum use may not have been made of the studio equipment and occasionally recordings have been poorly engineered. The fun part is the creative processing to increase the punch, impact, and energy or sweeten the sound. That final touch that can make the recording more enjoyable to listen to."

Ray Staff (AIR Mastering)

At the end of the mixing session, the studio will provide the final mix in whatever format is required (stereo, 5.1) ready for moving to the next stage of the music production line. Some mix engineers still prefer to use ½-inch analog tape for the analog factor, but this is happening less and less. DAT, or digital audio tape, is still considered a safe format but is dwindling fast from professional consciousness in light of other digital data formats such as hard drives, CDR, and DVD-Rs. It is most likely though that the music will leave the studio on one of these data formats in a .WAV file or similar. What now? Well, this music has one more creative path before it is all over: mastering.

Mastering is considered, to some, a dark art and is perhaps one of the most guarded secrets in modern music engineering. Mastering engineers spend time applying seemingly small amounts of processing with sometimes enormous results on the music. These engineers enjoy a "dark art" status, as the guarded secret of mastering is kept behind closed doors in a cloak of mystery. It is quite ironic that out of all the engineering processes a release goes though, mastering might apply the very least amount of compression or equalization. However, the impact and impression the mastering often provides can drastically enhance

What is Music Production. DOI: 10.1016/B978-0-240-81126-0.00012-3

the music. It is this that the engineer is paid for—the expertise, even hand, objectivity, and seal of approval.

However, this additional creative process has been developed over the years and is an important, critical aspect of releasing music. Understanding what has come before will allow you to appreciate what mastering is trying to achieve and why it remains a functional, creative and powerful tool in the music production process.

MASTERING WAS...

Well, *is* a process called pre-mastering, to be exact. Historically speaking, the mastering stage as we understand it today was not to necessarily improve the sonic characteristics of the material. Its main job was to match the material for the format it was to be duplicated to. For example, if the music was to be pressed onto vinyl, as most records were for so many years, the mastering engineer would need to equalize the work for that medium, ensuring that the bass was managed and the level would also need to be managed so that signals did not break through walls to the next groove when cutting, for example.

These skills are still out there and many of the major mastering houses still retain their lathes, or disc cutters, for this exact purpose. However, mastering as a purpose has shifted over the years gradually to a process that not only ensures that it makes the medium play accurately, but one that improves and polishes the sonic qualities of the music.

A HISTORICAL PERSPECTIVE

In AIR Mastering's Ray Staff's words: "In the early years, disc cutting engineers often worked with a minimal amount of equalization or compression. They were essentially transfer engineers. Many cutting engineers believed that the mix should be cut as flat as possible and the producer and artist were sometimes not even allowed to attend the cut. This prompted a lot of effort from some recording engineers to ensure their work was right when it left the studio. The idea that it can be fixed in the cut simply did not exist. Many engineers who trained at major studios had to spend a period of time cutting records before they were allowed to progress to recording engineer status, and this was invaluable experience for them. From around the end of the '60s mastering became far more creative. Independent cutting rooms started to provide a creative mastering environment. As each new format has arrived (cassette, 8-track cartridge, CD, and compressed audio for downloads) we have had to evolve into the high-quality mastering services you see today."

MASTERING IS...

Mastering is now considered the last creative stop on the train line to the end of a production project. While it is far from the end of the line, it is indeed the last station stop at which the audible content can be altered, improved, shaped, and prepared for market. As we have already established, mastering is often referred to as a Jedi art of audio engineering because of its ability to improve and manipulate the sonic quality of material with relative ease. It is considered to be a secretive art form for good reason. Many recording and mix engineers have the same equipment (more or less) and the same ability to equalize, compress, limit, and edit. This is not disputed, as these skills are more or less transferable. The art is in the skill, not the equipment.

MASTERING IS... IN PART, LISTENING

Where the mastering engineer differs from that of recording and mixing engineers is in the type of listening skills. As has been discussed earlier in this book, there are overarching listening types to note: macro listening and micro listening, in addition to holistic.

As a quick reminder, we established that micro listening is the ability to work with small elements of the mix, as a mixing engineer would, and scrutinize each element in fine detail, perhaps soloing the part in question, taking into account each instrument, its musical content, and sonic reproduction. Micro listening is essential for all engineers in all walks of life; however some will do this more often than others and for different reasons. A recording engineer will focus in on the instrumentation being recorded in order to ascertain that it has been played properly, it is in tune, and the best performance has been captured. The mastering engineer, on the other hand, will use this ability (listening to an already completed stereo mix) to focus in on elements within the mix that need scrutiny, such as a sibilant vocal, an errant noise, or anything else, for that matter.

Macro listening is the ability to objectively view the whole broadband audible content as one. To listen in this overarching way is an excellent skill that mix engineers and mastering engineers perhaps focus on that little bit more as the "whole" is their product. Listening to the whole album as one project could be considered *holistic* listening.

Listening in this way is a form of detachment that is important for the mastering engineer to learn so that an even hand can be applied to the music.

The secretive art mantle has been levied because of this detachment and the mastering engineer's ability to preside over the holistic, the album, and consider it as one complete entity. Many see the mastering process as quality control, ensuring that their mixes are in order, and to gain comments from those whose opinion they might trust that it is indeed "a good mix."

WHAT CAN BE DONE?

From the producer's perspective it is likely that you'll already mostly understand the available processes that your music can be put through, but as we've already established, these will be managed and attacked in very different ways. Many producers tend to rely on their chosen few engineers to master their work, choosing each one on their strengths depending on the style of music they are trying to tackle.

As we know, a number of processes are applied to each track to potentially lift its audible qualities to the best possible standards. A mastering engineer will have preferred tools and techniques that bring out the best in a piece of recorded material. Many consider this as *sweetening* of the music through the use of EQ and other tools such as dynamics processing.

Simple processing such as this may not only improve each track, but should be utilized to improve each track in relation to the rest of the album and ensure parity. Therefore the quality of the sound is retained across the whole work, as well as being ordered in such a way that it retains listener interest.

Therefore, the mastering engineer will sequence, or assemble the audio into an order and shape that makes it flow when played on a CD player. Bob Katz discusses in his book *Mastering, The Art and The Science* (Focal Press, 2007) the importance that the order of the music places on the flow of an album.

What the mastering engineer does exactly might not be immediately obvious, as each one manages his workflow in a very different way. The ideal is for a mastering engineer to take your mixes and to literally approve them without doing very much at all. In some cases this does happen and the mastering engineer is so impressed with the depth and clarity of the mixes that their work will be little. This is rare though.

Many a mastering engineer will have been given, at one point or another, an album that they have been asked to put their stamp on (or someone else's stamp; perhaps the label). Many albums are expected to be extremely loud, as we'll learn later, and in those instances, perfectly good, and dynamically presented mixes can indeed be squashed.

There are mastering engineers that work alone, and are happy for you to send the material over the Internet to them and they'll send some proofs back. This is known as eMastering. This has become very popular in recent years as it is usually a lower cost (as the engineer can work in their own antisocial time). This is a good option for producers wishing to save money in the current climate.

However, many producers and engineers (and of course artists) still like to attend sessions in the mastering studio and will pay the extra for it. These sessions ensure that the mastering engineer has live feedback on their work and can respond accordingly having serious discussions regarding the work.

Mastering can do many things to the audio from simply sweetening the audio with a bit of EQ, through to some more forensic style trickery to improve stereo image, or to dynamically compensate for a sibilant vocal or a ringy snare drum within the mix.

Many mastering engineers like to keep their options open, often suggesting that the mix engineer prepare a number of grouped tracks for the mastering session, which we'll look into next.

PREPARING YOUR MUSIC FOR MASTERING

Mastering engineers are understanding, even sympathetic, to both sides of the loudness coin and as such may request that your music comes to them with as much dynamic range as you feel you can allow. We understand the mixes need to be representative, but would prefer that no mastering compression be applied at the mix stage. Some engineers will insist on using buss compression to really allow them, and the A&R person at the back of the room, to gain a feel for what it will sound like post-mastering.

Some mixes go to mastering engineers with compression on. In some cases these are very good, but at other times this can cause more problems as they may need to unpick this compression to get the results you want over the whole album. Some mastering houses would prefer for a mix to come to them uncompressed, but the mix sounding excellent. They can then compress and limit your music using their mastering-grade techniques to obtain the best results. Alternatively, perhaps provide them with both the compressed version and a version without.

You might be required to send the material to the mastering house using the Internet. Most audio files these days are digital, even if they are of a much higher sample frequency and bit depth than years gone by. So this can be sent over reliably and quickly to the other side of earth for mastering if necessary.

You could simply send the mastering house a stereo file of the mix. However, there are some other alternatives such as stem mixing. There are many ways in which this can be broken down. Essentially stem mixing means that you provide the mastering engineer with the mix, including the mixing engineer's EQ, compression, and reverb, in stereo pairs. So in addition to the mix file, the stems might be as follows:

- Drums
- Bass
- Guitars
- Keys
- Backing vocals
- Vocal

Or a pared down version might be more commonly:

- Your mix
- Main vocal (no backing vocals)
- No main vocals (with backing vocals)
- No vocals at all (instrumental)

Stem mixing, like the second example, allows the mastering engineer to take the true flavor of the mix engineer's vision and work and can compensate a little bit with the vocals or by adding more of the vocals into the mix (as it's all phase-aligned audio in a mastering-ready DAW). Alternatively if it's backing that's needed she could mix in the "no main vocals" one to make the track more punchy. Other options include treating one stereo pair differently to another using EQ and compression, or duplicating tracks to employ parallel compression.

Considering how you are going to present your material to the mastering process will dictate how many different mixes of each track you need to remember to bounce at mixing.

Once the mastering engineer has what she needs, she'll crack on and produce the masters for you, but in the interim she'll need the codes. The all important codes.

THE CODES

PQ Codes

At the mastering suite, the engineer will also be very concerned about logging and details. Mastering is not only a creative role, it also encompasses some considerable administration. It is the engineer's duty to provide the manufacturer with what is known as a PQ sheet, or listing. These lists detail the P and Q flags in the CD data. The P flag simply informs the CD player that a new track is following, while the Q flag contains data about the track's length and duration and codes that give the track its identity.

ISRC (International Standard Recording Code)

The International Standard Recording Code was developed to sit beside each digitally encoded audio track as a means of identification. It has many purposes, but the main one is for automation of royalty payments. Every time a song is played on the radio, systems are available to log that the track has been played and, in turn, the appropriate royalty payment is transferred to the appropriate collection agencies.

Prior to the mastering session, it is advisable to ensure that all ISRC codes are obtained well in advance so they can be encoded and added to the PQ sheet.

UPC/EAN (the barcode)

The Universal Product Code (UPC) was originally introduced in the 1970s to log products as they left the grocery store in America. The EAN (European Article Number) is the European equivalent. The record label will normally supply this code. As with the ISRC, it is prudent to ensure that you have this code well in advance of the mastering session.

Other codes

Other codes are often required at the session along with other details, such as any additional catalog or reference codes. Other information such as the record label or client will also be required at times.

Some of the main codes that are often asked about are things such as CD text. CD text can be programmed, in a similar way to the user programming information about the track into a mini-disk machine. CD text will show up on many CD players and is sometimes used, although it is less common these days.

```
//////////////////////////////////////////////////////////////////////
//
//    Table of Content generated by Pyramix Virtual Studio CD Mastering System
//
//////////////////////////////////////////////////////////////////////


Disc Title    : Guitar Revival 2
Label         : Universal Publishing Production Music
Date          : 15 March 2008

Customer Name    : Universal Publishing Production Music
Customer Contact : Joe Bloggs
Customer Phone   : 020712345678

Master ID Code :
Ref. Code      : C07-20
UPCEAN Code    : 0000000000000

Track #   Index #   Time       ISRC/Name          Copy
----------------------------------------------------------
01        00        00:00:00   GBAZC0834301  yes
          01        00:02:00   Sabotage
Length              02:44:56
----------------------------------------------------------
02        00        02:46:56   GBAZC0834302  yes
          01        02:48:34   East Bay Livin'
Length              01:56:74
----------------------------------------------------------
03        00        04:45:33   GBAZC0834303  yes
          01        04:46:64   Tight Jeans & Trilby Hats
Length              02:54:03
----------------------------------------------------------
04        00        07:40:67   GBAZC0834304  yes
          01        07:42:71   Time to Shout
Length              02:41:43
----------------------------------------------------------
05        00        10:24:39   GBAZC0834305  yes
          01        10:27:53   Arctic Melt
Length              02:56:07
----------------------------------------------------------
06        00        13:23:60   GBAZC0834306  yes
          01        13:25:16   Jump Start
Length              02:33:05
----------------------------------------------------------
07        00        15:58:21   GBAZC0834307  yes
          01        16:00:00   Lazy Eye
Length              03:24:17
----------------------------------------------------------
08        00        19:24:17   GBAZC0834308  yes
          01        19:27:24   Take Me Out Tonight
Length              02:46:00
----------------------------------------------------------
09        00        22:13:24   GBAZC0834309  yes
          01        22:14:36   Time On Our Side
Length              02:56:33
----------------------------------------------------------
10        00        25:10:69   GBAZC0834310  yes
          01        25:12:56   Black Roses
Length              03:06:67
----------------------------------------------------------
11        00        28:19:48   GBAZC0834311  yes
          01        28:21:27   The Night Is Right
Length              02:39:09
----------------------------------------------------------
12        00        31:00:36   GBAZC0834312  yes
          01        31:01:71   Sidewalk
Length              02:55:48
----------------------------------------------------------
13        00        33:57:44   GBAZC0834313  yes
          01        33:59:35   Arctic Melt (Acoustic)
Length              02:17:34
----------------------------------------------------------
14        00        36:16:69   GBAZC0834314  yes
          01        36:19:21   Lazy Eye (Acoustic)
Length              03:28:30
----------------------------------------------------------
15        00        39:47:51   GBAZC0834315  yes
          01        39:54:45   Time On Our Side (Acoustic)
Length              02:57:18
----------------------------------------------------------
16        00        42:51:63   GBAZC0834316  yes
          01        42:55:10   Sidewalk (Acoustic)
Length              02:47:05
----------------------------------------------------------
17        00        45:42:15   GBAZC0834317  yes


          01        45:43:73   Sabotage 30
Length              00:30:00
----------------------------------------------------------
18        00        46:13:73   GBAZC0834318  yes
          01        46:15:55   East Bay Livin' 30
Length              00:30:00
----------------------------------------------------------
19        00        46:45:55   GBAZC0834319  yes
          01        46:47:37   Tight Jeans & Trilby Hats 30
Length              00:30:00
----------------------------------------------------------
20        00        47:17:37   GBAZC0834320  yes
          01        47:19:19   Time To Shout 30
Length              00:30:00
----------------------------------------------------------
21        00        47:49:19   GBAZC0834321  yes
          01        47:51:01   Arctic Melt 30
Length              00:30:00
----------------------------------------------------------
22        00        48:21:01   GBAZC0834322  yes
          01        48:22:58   Jump Start 30
Length              00:30:00
----------------------------------------------------------
23        00        48:52:58   GBAZC0834323  yes
          01        48:54:41   Lazy Eye 30
Length              00:30:00
----------------------------------------------------------
24        00        49:24:41   GBAZC0834324  yes
          01        49:26:23   Take Me Out Tonight 30
Length              00:30:00
----------------------------------------------------------
25        00        49:56:23   GBAZC0834325  yes
          01        49:58:05   Time On Our Side 30
Length              00:30:00
----------------------------------------------------------
26        00        50:28:05   GBAZC0834326  yes
          01        50:29:62   Black Roses 30
Length              00:30:00
----------------------------------------------------------
27        00        50:59:62   GBAZC0834327  yes
          01        51:01:44   The Night Is Right 30
Length              00:30:00
----------------------------------------------------------
28        00        51:31:44   GBAZC0834328  yes
          01        51:33:26   Sidewalk 30
Length              00:30:00
----------------------------------------------------------
29        00        52:03:26   GBAZC0834329  yes
          01        52:05:08   Arctic Melt (Acoustic) 30
Length              00:30:00
----------------------------------------------------------
30        00        52:35:08   GBAZC0834330  yes
          01        52:36:65   Lazy Eye (Acoustic) 30
Length              00:30:00
----------------------------------------------------------
31        00        53:06:65   GBAZC0834331  yes
          01        53:08:47   Time On Our Side (Acoustic) 30
Length              00:30:00
----------------------------------------------------------
32        00        53:38:47   GBAZC0834332  yes
          01        53:40:29   Sidewalk (Acoustic) 30
Length              00:30:00
----------------------------------------------------------
AA        01        54:10:29
```

Part of an example PQ sheet showing all the relevant codes used such as ISRC.

DELIVERY
To the mastering house

For the delivery of audio material, mastering houses are ever more commonly working remotely from the production team, receiving files over the Internet using FTP (File Transfer Protocol). In the same manner, mastering engineers can also post their completed masters to the manufacturers using FTP too.

Lots of audio is still brought to the mastering house in the same way it always has via analog tape, DAT, CDR, and other data formats. One thing that must be mentioned at this stage is labeling. Make sure that your labeling is something you understand and can be conveyed easily to the mastering house. Which file is the master? Which is the copy? Which version are we supposed to be using? Clarity on such issues will save time and lower costs in the long run.

Mastering engineers will, of course, vary, but most will prefer you to send them the highest resolution file you can. Certainly 24 bit and as high a sample rate as the recording was made on. Typically this is often 44.1 kHz and 24 bit. However, higher sample rates are often used and should be passed to the mastering house. Leave it with the mastering engineer to get it to 16 bit 44.1 kHz CD-ready audio.

To the manufacturer

Mastering houses will produce the manufacture-ready master in a range of formats. Most typically this will be a DDPi (Disc Description Protocol image file). These are a collection of data files that continue to be the most reliable. With a DDP, it will either work or it won't. DDP can either be stored on a CDR (if it will fit), a DVD-R, hard drive, Exabyte tape, or uploaded to an FTP server. Manufacturers accept audio CDs as the master for duplication, although this is not advisable given the opportunity for error on the disc. Ensure the PQ sheet with all the completed information tallies to the audio sent.

Some mastering engineers such as Ray Staff are producing enhanced CDs for their clients that include a data section. This data section holds some MP3s of the album's material that has been data compressed at the mastering house using specially prepared masters. These masters have been adjusted especially for the MP3 player market and with the perceptual coding systems used in mind, thus providing superior quality playback.

RAY STAFF ON FORMATS

There have been many arguments over the years about the varying quality of manufactured CDs. I would suggest there are three main areas at which the quality can be influenced: the CD master, the glass mastering, and the quality of the actual pressing.

There are now two main mastering formats available to us to send to the manufacturers. They are CD audio and DDP. The format used for your CD master is important for the quality of the final product and its storage as part of a record companies assets.

The CD is easy, convenient and supported by all good DAW systems. It is essential to have a good quality CD writer running at the optimum speed (x1 is the speed of choice for many mastering engineers) and to have high-quality CD blanks. Once written, the CD should be evaluated for disc errors using a CD analyzer and auditioned for content errors.

There is also some confusion about the terms PMCD and Red Book CD. Red Book is the original specification laid down by Sony and Philips for the Audio CD (CDDA). PMCD is a format that was introduced by Sony, Sonic Solutions, and Start Lab. There is a small additional piece of hidden data containing the "PQ or PreMaster Cue Sheet ." Only Sonic Solutions systems can write a PMCD and only Laser Beam Recorders manufactured by Sony can read them. Essentially both offer the same quality.

I personally find the quality variations of this format unreliable. It takes a well-configured system with carefully selected blanks to produce good results. The quality of cloned copies is also very variable and I have often seen this type of CD master physically deteriorate, reducing its reliability for long-term storage and future use.

DDP (Disc Description Protocol, developed by Doug Carson & Associates) is a method of storing an audio CD as data. Only the more professional DAW systems traditionally included this option. DDP masters were originally written to an Exabyte tape. As a collection of files they can be stored on a hard drive (DDPi or DDP image file) and shipped via a ROM disc or via an FTP site. This format offers reduced delivery costs and speeds, high quality, and efficient long-term storage. Many large record companies use DDP files to archive their product as they can be cloned repeatedly (using good error-checking software) with no audible deterioration. Using large data archiving and media asset management systems, they can have an infinite storage life and high access speeds for future manufacturing. Again, quality control is essential and a CD should be made from the DDP master and auditioned for content errors before dispatch to the plant.

Most online music stores will only accept an audio CD as the source for making the MP3s on their site. It would be a positive step forward if they too would accept a DDP master. With difficulty, we have sometimes persuaded providers to use WAV files. The quality of many online or homemade lossy data compression files varies considerably. To this end some mastering rooms do offer a service to make high-quality files. On a number of projects we have been asked to add these to the data area of an Enhanced CD to give the quality conscious consumer an alternative.

THE LOUDNESS WARS

Much of the mastering engineer's skill has been abused over recent years to develop what some in the industry have dubbed *The loudness wars*. Producers, engineers and mastering engineers all agree that levels have increased so much that they have literally hit the red!

As an example of the issue, I must raise the tone of this text to a scene from Spinal Tap. Nigel Tufnel (played by Christopher Guest) is showing their "rockumentary" maker Marty DiBergi (Rob Reiner) his collection of guitars. Nigel proudly points out his custom guitar amp on which the volume control markings go up to 11 and not the standard 10. Nigel proudly states that "when you get to 10, where can you go? Nowhere!... this goes up to 11!". Rightly Marty suggests "Why don't you just make 10 louder?' A dazed and confused Tufnel retorts after a long, painfully confused yet thoughtful pause, "This goes one louder" (*This is Spinal Tap*, 1984). In all seriousness, though, we've come to a pivotal crossroads in the production of modern music that many audio professionals are concerned about.

This Spinal Tap analogy actually works. The maximum loudness has been achieved by the system we have in place—digital audio. The only way we can go is down, as we cannot make "10 louder" without accepting a new digital standard. Digital audio will always have an upper limit of 0 db in its current format and this is the level which we're all aiming for and nothing less will seemingly do (see Ray Staff's sidebar Loudness, below). However, at some point we have to say "no more," as audio quality is suffering overall despite some valiant efforts by the mastering community to make their masters still shine through these restrictions.

Recently, having had the fortune to listen to the direct mixes of a rated album, we were astonished by the post-mastered results. The album was produced by an A-list producer and mixed by an exceptional team. The mixes were dynamic, wide, with good use of frequency range and simply beautifully crafted. What happened at the mastering stage? The released album is distorted in some way at loud sections and does not translate very well on some excellent monitoring systems. Was the mastering engineer asked to raise the levels by inexperienced record label executives, or the artist(s) themselves?

RAY STAFF ON LOUDNESS

Digital recording and mastering levels have increased dramatically in recent years. Most digital equipment is supplied with digital peak reading meters only with a scale from 0 dB downwards. The manufacturers were keen that we should avoid the distortion caused by digital clipping. Unfortunately, the digital zero has become a target and people assume that if the meter is not hitting the highest possible level, their recording will be deficient in some way and will not sound very loud.

The only meter in common use that gives any true indication of loudness is a VU (Volume Unit) meter. To maintain a consistent level across a number of songs they should all have a similar VU reading. The final relative level adjustments must be made by ear. Peaks within these tracks may vary considerably depending on the musical content. When digital was first introduced, most studios would set their analog console 0 VU to equal −14 on the digital scale, leaving a respectable amount of headroom for transients. Many CDs are now so hot it has become necessary to calibrate 0 VU to equal −4dB on the digital scale. That means if the average level is 0 VU there is only about 4 dB left for transients and peaks.

Some remix rooms are adding large amounts of compression and limiting just because they believe it is the right thing to do, with little consideration given to the resultant sound. All too often this is mixed within a computer that is struggling with the number crunching, using cheap poorly adjusted plugins. In the mastering room we are told, "I think my mix sounds a bit small, but we want our CD to sound big and loud." Unfortunately, the damage has been done. Too much or the wrong type of dynamic processing can make music sound small with a lack of depth and punch. This is, of course, the worst possible scenario and fortunately there are many good mixes out there.

It is not wrong to use compression as it is an invaluable tool, but I really think that studios should check the effectiveness of any compression or limiting used in the mix. When using any audio processing that adds gain, it is important to compare it to the original with the level matched. Only then can you really determine if it is doing a worthwhile job. If in doubt, don't do it. That gives the mastering engineer the opportunity to work with a recording that has got potential to be manipulated and worked into a great sounding CD. I am pleased that some DAW systems now offer the ability to use a VU scale alongside a peak reading meter. This now gives the operator the ability to have a better judgment of the loudness.

The mastering should be auditioned at different levels to ascertain the best possible result and consideration given to what happens after the mastering stage. Lossy media files such as MP3 may distort if the level is too hot and it is a fallacy that a CD has to be loud to sound good on the radio.

I am a keen supporter of the Bob Katz K-System (Katz, B. 2007. *Mastering Audio: The Art and The Science.* Focal Press). This is an extension of the original alignment method I described above and also links the recording level to an acoustic reference level. It's a good common-sense approach to engineering and removes digital zero target.

WHAT IS LOUDNESS?

Loudness is one of those terms that are rather unquantifiable. Yes, we do have all manner of units of measurement for loudness in one form of energy or another. However, what they mean in each case can be interpreted in different ways. For example, in audio engineering we have several ways of explaining how loud something is and each has its own part in the engineering chain. Terms such as *gain, level, volume, intensity, power,* and *loudness* all refer to how loud we think something is. These all have different functions and these differences are not addressed here, but outline some of the problem.

As Ray Staff has pointed out, 0 dB on a modern peak meter has become a target as opposed to a warning. Peaks of music will be hitting this at all costs at the request of many clients. However, it is not the peaks that represent the loudness, but the RMS, or average power, of the signal. This is best represented by what we call a Volume Unit (VU) meter as shown on page 252.

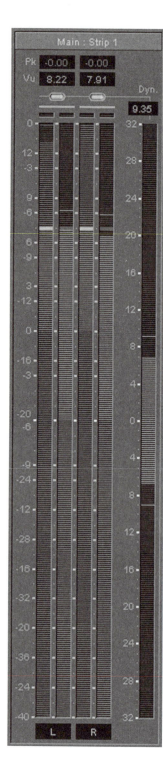

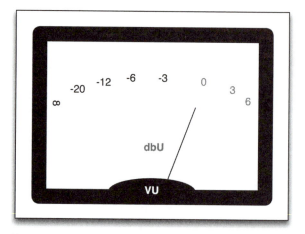

Merging Pyramix stereo meters can show both peak meters (the louder of the ones shown) and VU meters (the lower), plus a useful dynamics meter (to the left) showing the difference between the peak and VU levels. In this example, the program material is very loud showing an average level of around −6dB below and peaking at 0 dB FS.

The problem of loudness in some modern records is exacerbated because records are competing to be louder than each other. Making every record louder than the last means that the natural dynamics of the next audio material has to be less than the one before to fit into the set amplitude scale of digital audio.

The K-system

The management of dynamic range is an important one that Bob Katz addresses in his book *Mastering Audio: The Art and The Science* (2007). He outlines a metering system ("tied to monitoring gain" so that the perception of loudness is not altered) that proposes that the loudness of 0 VU is regulated for the type of activity, whether that be cinema, music, and so on. In each K-System, 0 VU is intended to be linked to 83 dBC SPL (for further details of this, please see Bob Katz's book).

For most stages such as recording and mixing, Katz proposes that this is set to −20 below full scale, allowing plenty of headroom for the peaks— K-20 as it is known in the K-System. Headroom is the space available in the amplitude range that allows for dynamic range. For mastering he proposes we use "K-14 for a calibrated mastering suite."

In such a proposal, 0 VU would be the target as opposed to the full scale of the dynamic range. This would immediately improve the quality of the music at all points in the chain and at the same time allow for the natural dynamics that make music breathe.

In conclusion, mastering is a key process to ensure your production sounds and feels like the professional job it should. Choosing your mastering engineer and mastering house will be to personal taste. Many producers will choose to attend the session, but equally many are happy these days to send the masters to their favorite engineer to get straight on with it using what is now known as eMastering.

SECTION E
The Future

CHAPTER E-1
The Changing Face of Music Production

Speaking about the DIY culture and artist development, Tommy D explains that producers should promote that the new way of working is "alright. You'll be ok. It is a bit scary I know. It's a lot more work, but you know what? It's fun! It's just as much fun as sitting in a studio telling a guitarist what to play… the building of an artist is different but just as fun."

What a journey we've had. We sat together one rainy spring morning in Starbucks in Leeds, U.K. while our colleagues at Leeds College of Music were marking the students' final dissertations for their Degrees and Masters qualifications. We took some time out to analyze whether our first-year degree students had 'got it' after one year of tuition in music production. While they definitely understood the basic processes, equipment, and so on, they did not yet have an appreciation of what the music producer did.

This, we concluded, was not because of the program of study. Neither was it because they lacked any access to fantastic industry personnel who gave their time to come and tell students how it was. We wondered whether it was because every book on the reading lists might be focusing on the equipment, and perhaps showing music production in an unrealistic and glorified light.

With lattes and Americanos in hands we set about, there and then, to write the proposal and contents to a book we hoped might change that perception for students in the future. Those coffees were drunk over four years ago now and the book has gone through considerable changes, just as the industry has.

However, we have been fortunate to have had the pleasure of discussing our favorite topic with many experienced producers, engineers, and music industry professionals from around the world. We hope the image we have produced of the industry is fair, representative, and yet still incredibly positive. The future is, as one producer stated, "a little frightening, but we're an innovative bunch." The positivity we felt from our interviews made us more inspired and confident for our students.

What is Music Production. DOI: 10.1016/B978-0-240-81126-0.00013-5

How did we gather this sentiment? Well, in many of the interviews we held we asked what the future of the music industry was going to be like. Many suggested that it was simply going to be different. Different in that the mode of business is shifting. Shifting from the label-led dominance, to a much more do-it-yourself culture. Many of the producers we spoke with had taken it upon themselves to find and develop their own artists, often at their own expense, and often with mixed success.

The future is going to be interesting. An open opportunity for marketing new music now exists for all with the right product and the ability to somehow create a buzz. As we discovered in this book, there are many new tools to assist the producer to promote their acts and develop artists in the more holistic sense rather than just in the studio.

Most producers did agree that the industry as we once knew it has largely declined. The big advances from the labels, the long stints in a residential studio, and so on, have for most subsided. However, the advances in technology mean that sensible new working practices and efficiency measures have come to the fore.

Professionals can now tool up to work creatively and productively at home, offering greater value of time and dedication to get things right, and on far reduced budgets. We conclude from this book that the income streams have changed and that to develop material with artists has become as common as has the 360° deal. New ways of working for a new musical world are with us.

Succeeding in this industry, at whatever level you aim for, will take dedication, inspiration, and simple old-fashioned hard work. We hope this book shows that not all the skills required are in studio wizardry, musical prowess, or music business alone. We hope that this guide to guidance in music production shows the producer as a highly rounded, talented, and broadly skilled individual who will adapt and drive the industry forward.

SECTION F
Appendices

The Tape Store

RECOMMENDED READING AND LINKS

The recommended reading is provided on a chapter-by-chapter basis. This provides you the opportunity to gen up much more in each subject area as you wish. These are by no way exhaustive, but should prove a good signposting tool.

SECTION A – WHAT IS MUSIC PRODUCTION?

Burgess, R. J. (2005). *The Art of Music Production*. Omnibus Press.

Eno, B. (1996). *A Year with Swollen Appendices: Brian Eno's Diary*. Faber & Faber.

A-1 Quantifying It

Cunningham, M. (1998). *Good Vibrations: A History of Record Production*. Sanctuary Press.

Hepworth-Sawyer, R. (Ed.), (2009). *From Demo to Delivery: The Process of Production*. Focal Press.

Massey, H. (2000). *Behind The Glass*. Backbeat Books.

Massey, H. (2009). *Behind The Glass* (Vol. II). Backbeat Books.

Emerick, G., & Massey, H. (2006). *Here, There and Everywhere: My Life Recording the Music of The Beatles*. Gotham Books.

A-2 Analyzing It

Moylan, W. (2007). *The Art of Recording: Understanding and Crafting The Mix*. Focal Press.

Bartlett, B., & Bartlett, J. (2005). *Practical Recording Techniques*. Focal Press.

Watkinson, J. (2001). *The MPEG Handbook*. Focal Press.

SECTION B – BEING IT

B-1 Being a Producer

Kintish, W. (2008). *I Hate Networking*. Jam Publications.

Harding, P. (2009). *PWL From The Factory Floor*. W.B. Publishing.

Daylite CRM. www.marketcircle.com.

Sage ACT CRM software. www.sage.co.uk/act.

Microsoft Dynamics CRM. www.crm.dynamics.com.

SalesForce CRM. www.salesforce.com.

Filemaker. www.filemaker.com.

Altermedia's StudioSuite. www.studiosuite.net.

Farmers WIFE. www.farmerswife.com.

B-3 Being a Business

Allen, D. (2001). *Getting Things Done*. Piatkus.

Forster, M. (2006). *Do It Tomorrow and Other Secrets of Time Management*. Hodder & Stoughton.

Stone, C. (2000). *Audio Recording for Profit: The Sound of Money*. Focal Press.

What is Music Production. DOI: 10.1016/B978-0-240-81126-0.00016-0

SECTION C – PREPPING IT:

C-1 What's the Deal?

Harrison, A. (2008). *Music, The Business*. Virgin Books.

Passman, D. S. (2008). *All You Need to Know About the Music Business*. Penguin.

Avalon, M. (2009). *Confessions of a Record Producer: How to Survive the Scams and Shams of the Music Business*. Miller Freeman Books.

C-2 Pre-Production

Huber, D., & Runstein, R. (2005). *Modern Recording Techniques*. Focal Press.

Cope, D. (2009). *Righting Wrongs in Writing Songs*. Cengage Learning.

Teaboy: An online repository for your recall settings for outboard. www.teaboyaudio.com.

C-3 Project Management

Association for Project Management. (2006). *The APM Body of Knowledge*. APM Knowledge.

C-4 The Desired Outcome: Strategies for Success

Burgess, R. J. (2005). *The Art of Music Production*. Omnibus Press.

Farinella, D. J. (2006). *Producing Hit Records: Secrets from the Studio*. Schirmer Trade Books.

Harding, P. (2009). *PWL from the Factory Floor*. WB Publishing.

SECTION D – DOING IT

Burgess, R. J. (2005). *The Art of Music Production*. Omnibus Press.

Farinella, D. J. (2006). *Producing Hit Records: Secrets from the Studio*. Schirmer.

D-1 The Session

Huber, D., & Runstein, R. (2005). *Modern Recording Techniques*. Focal Press.

D-2 The Mix

Izhaki, R. (2008). *Mixing Audio- Concepts, Practices and Tools*. Focal Press.

Stavrou, M. P. (2003). *Mixing with Your Mind*. Flux Research Pty.

D-3 The Mastering Session

Katz, B. (2007). *Mastering Audio: The Art and The Science*. Focal Press.

Cousins, M., & Hepworth-Sawyer, R. (2010). *Logic Pro 9: Audio & Music Production*. Focal Press.

Bob Katz's website. www.digido.com.

Russ Hepworth-Sawyer's website. www.mottosound.co.uk.

International Standard Recording Code (ISRC). www.ifpi.org/content/section_resources/isrc.html.

WEBSITES

The most important website is *www.whatismusicproduction.com* associated to this site.

Trade Associations & Links

PRODUCTION & ENGINEERING

U.K.

www.mpg.org. The Music Producers Guild. A body representing music production professionals in the U.K.

www.jamesonline.org.uk. Joint Audio Media Education Services. A joint venture between the Music Producers Guild and the Association of Professional Recording Services supporting education through accreditation.

www.aprs.co.uk. The Association of Professional Recording Services.

www.britishacademy.com. British Academy of Composers and Songwriters.

USA

www2.grammy.com/recording_academy/producers_and_engineers/. The Producers and Engineers Wing (The P&E Wing) of the Grammys.

www.aes.org. The Audio Engineering Society.

www.spars.com. Society of Professional Audio Recording Services.

MUSIC AND MUSIC MANAGEMENT

www.musiciansunion.org.uk. The Musicians Union in the U.K.

www.themmf.net. The Music Managers Forum.

www.afm.org. The American Federation of Musicians.

RECORD INDUSTRY

www.bpi.co.uk. The British Phonographic Industry (U.K.).

www.riaa.com. The Recording Industry Association of America.

Online Networks

www.prosoundweb.com. A large community to discuss all matters relating to audio.

www.myaprs.co.uk. MyAPRS: APRS' online community.

www.gearslutz.com. Gearslutz: Another large online audio community.

Audio Magazines

www.prosoundnewseurope.com. An excellent audio magazine for the European industry with fantastic equipment reviews.

www.prosoundnews.com. Audio magazine for the American industry.

www.eqmag.com. EQ magazine is a good source of information for the recording musician.

www.musictechmag.co.uk. A U.K. magazine for producers and recording musicians.

www.soundonsound.com. A long-running U.K. music technology magazine.

www.resolutionmag.com. A detailed U.K. audio magazine.

www.musicweek.com. U.K.'s leading music business magazine.

www.billboard.biz. U.S. music industry news.

Royalty Collection Agencies

U.K.

www.prsformusic.com/Pages/default.aspx. PRS for Music is an alliance of the Performing Rights Society (PRS) and the Mechanical Copyright Protection Society. A U.K.-based collection agency for the writers of recorded music.

www.ppluk.com. PPL, or Phonographic Performance Limited. PLL and VPL (Video Performance Limited) collect royalties for the performance of recorded music (or music videos in the case of VPL) on radio, television, etc.

U.S.A.

www.soundexchange.com. Sound Exchange is a new performance rights organization (PRO) in the United States primarily concerned with collecting digital performance royalties.

www.ascap.com/index.aspx. The American Society of Composers, Authors and Publishers. ASCAP collects royalties on behalf of musical copyrights just in the same way that PRS does in the U.K.

www.bmi.com. Broadcast Music Inc., like ASCAP, collects royalties for their composers and publishers.

www.sesac.com. Society of European Stage Authors and Composers—no longer just collecting for Europeans, is based in Nashville, U.S.A. and unlike most PROs is a for-profit company.

CANADA

www.socan.ca. SOCAN, Canada (Society of Composers, Authors and Music Publishers of Canada).

REST OF THE WORLD

Some of the main territories provided here.

www.jasrac.or.jp. JASRAC, Japan (Japanese Society for Rights of Authors, Composers and Publishers).

www.apra.com.au. , Australia.

www.sadaicmia.com/. APPASADAIC, Argentina.

www.cash.org.hk. CASH, Hong Kong.

www.rao.ru. RAO, Russia.

www.iprs.org. IPRS, India.

www.sacm.org.mx. SACM, Mexico.

www.apra-amcos.com.au. APRA|AMCOS, Australia.

EUROPE

www.imro.ie. IMRO, Ireland.

www.sacem.fr. SACEM, France (the Société des Auteurs, Compositeurs et Editeurs de Musique) is the PRS equivalent in France.

www.sabam.be. SABAM, Belgium (Société d'Auteurs Belge – BelgischeAuteurs Maatschappij) is a similar agency in Belgium but is not restricted to music, and represents film directors, scriptwriters, and poets among many other artists.

www.gema.de. GEMA, Germany (Gesellschaft für Musikalische Aufführungs- und Mechanische Vervielfältigungsrechte).

www.sgae.es. SGAE, Spain.

www.spautores.pt. SPA, Portugal.

www.aepi.gr. AEPI, Greece.

www.akm.co.at. AKM, Austria.

www.suisa.ch. SUISA, Switzerland.

www.bumastemra.nl. Buma Stemra, the Netherlands.

www.tono.no. TONO, Norway.

www.teosto.fi. TEOSTO, Finland.

www.stim.se. STIM, Sweden.

www.siae.it/. SIAE, Italy.

www.koda.dk. KODA, Denmark.

www.listir.is/stef. STEF, Iceland.

www.zaiks.org.pl. ZAIKS, Poland.

16 bit A common wordlength used for digital audio. The CD standard is 16 bit.

360 deal Deals created to encapsulate and draw from all aspects of an artist's income. A label might insist on a percentage of an artist's live merchandise income, for example.

44.1 kHz A common sample rate used for digital audio recording. The CD standard is 44.1 kHz.

AAC See "MP4."

Abbey Road Possibly one of the most famous studios in the world. Owned by EMI, this legendary studio in North London was where the Beatles recorded, among many others.

ADT Artificial Double Tracking, the process of doubling a track without the need for a second performance. Most commonly used on vocals and guitars in order to thicken the sound and texture. ADT can be achieved via effects processors or by manually copying the audio to a second track and delaying and modulating the copied audio. It was famously developed and used by George Martin and Geoff Emerick on John Lennon's vocals.

Advance When an artist or producer enters into a contract with a label, they may be provided with an advance of money. This advance, for producers, will often be for paying studio fees and so on in the case of *all-in* budgets.

AES Audio Engineering Society.

AES/EBU Collaboration between the Audio Engineering Society and The European Broadcast Union to develop a digital audio protocol used in professional audio applications.

AIFF Audio Interchange File Format, a non data-compressed audio file format common to the Apple Mac.

AIM The Association of Independent Music, a nonprofit trade organization for independent record companies and distributors in the U.K.

Akai Manufacturer of audio and hi-fi equipment. Akai is most renowned for the series of samplers, such as the S1000 and S1100.

Algorithm An algorithm is code, a little like a program, that achieves a certain task. For example, the reverberation algorithm for each setting on a digital effects unit.

All-in deal A type of deal made between a record label and a producer. The all-in deal involves the producer receiving a recording budget/lump sum out of which all recording costs (studio, musicians, engineer, etc.) and their own personal fee will be paid. This requires a high degree of budget management from the producer but gives greater control over how the money is distributed and spent. Any money that is left over from the initial sum can be kept by the producer on top of their personal production fee. This deal can therefore be beneficial to the producer depending on how efficiently the budget is managed.

Ambience The space around a recording. Can also be considered as a type of reverberation, or the depth in a mix and the atmosphere it can provide.

Analog A term used to describe audio recording formats prior to digital where the movement of a stylus was analogous to the waveform movement, for example.

APRS Association of Professional Recording Services.

Arrangement The organization of the musical structure and instrumentation used within a song or composition. The choice of instrumentation and its harmonic voicing forms a major part of an arrangement.

Arranger Someone brought in to orchestrate parts for session musicians, normally ensembles such as orchestras, chamber orchestras, and horn sections.

Artist & Repertoire Artist and Repertoire (more commonly known as A&R) is the department within a record label assigned with the duty of seeking out, recognizing, and signing talent.

Artist development The process of guiding the artist (also known as the talent) to become a unique artist. This development could be musical, stylistic, performance skills, and so on. Artist development can also refer to a department within a label charged with this function.

What is Music Production. DOI: 10.1016/B978-0-240-81126-0.00017-2

Assistant engineer In a traditional studio hierarchy the assistant engineer is considered to be one up from the tape op. This role involves a variety of duties in order to assist the main engineer on the recording. Among others these may include the setting up and recall of settings on the mixing desk, the organization of inputs and rerouting or patching, the patching in of outboard equipment such as effects units and dynamics processors, and so on, and the set up and positioning of microphones for tracking. If using analog tape, the assistant engineer or tape op may also set up the reel-to-reel machine ready for tracking.

Auratones A common nearfield monitor popular in the 1980s. A pair would adorn many a meterbridge. The Auratones were small speakers that imitated some of the lower cost audio equipment on the market.

Automation Refers to the ability for a mixing desk, or the Digital Audio Workstation equivalent, to be controlled remotely. This remote control is plotted against time so that levels can be automatically managed and maintained. Automation is now an integral part of modern mixing.

Automation modes Refers to the types of passes that can be managed when using an automation system. *Write* refers to a complete overwrite of any previously recorded data, a good place to start; *touch* refers to the ability for the faders to continue to track previously recorded automation but be updated when the fader is touched by the user; *latch* is where the fader can be moved and it does not resume its old automation as it would with *touch*.

Auxiliaries A control on a channel strip on the mixing console. Its function is to send a small submix out to allow some of the signal to be used either for an effects unit (typically post fader auxiliary) or for a headphone mix (pre-fader auxiliary).

Balance A term used in the studio to describe the mixing process. Getting the balance will be an engineer moving faders and making a balance of the recorded music. Balance can often also mean the difference in power from left to right like a pan control on one channel.

Board See "Mixing console."

Booth Refers to a soundproof small room that is acoustically treated to sound a certain way. Drum booths are common as they isolate the drummer's performance from spilling over the rest of the microphones in a live studio recording.

Bottom end A term used to describe the bass frequencies.

Buss compression Compression placed across a buss. An example could be a stereo compressor placed across a drum kit sent via a group buss.

Cans See "Headphones."

C.A.P. See "Producer's C.A.P."

Chorus (song section) The centrepiece of any song, which serves to deliver the main lyrical and melodic refrain or hook. The chorus is generally considered to provide the release point relative to other sections that may build tension within the song. The chorus provides the peaks in terms of the emotional architecture of a song or track. It is the chorus section that most songwriters and producers would want the listening public to remember.

Click track A track used to maintain strict tempo during the recording process. Sometimes used when tracking drum parts. The click can be generated from a number of sources, and can either be recorded to a track of its own as audio or can be sent as a MIDI signal to an external device such as a sound module.

Commit to tape A phrase still used today to describe recording something, even though open reel tape is not necessarily the most widespread format used in studios today. The expression still lives on.

Comping From the word "compiling." This abbreviated terminology refers to the compiling of various sections from various takes in order to compile one complete successful take. Can be used across many parts and instrumentation but most commonly used with vocals and drums, hence the expressions vocal comp or drum comping.

Compressor A device that reduces the dynamic range of audio signals.

Conductor An orchestra or ensemble of musicians looks toward the conductor for musical direction while performing an orchestrated piece.

Control room The room in the studio where the mixing console and studio equipment is housed.

Costs deal A type of deal or agreement made between a record label and a producer. The costs deal involves the producer receiving a production fee from the label for producing the songs or the album, but all recording costs (studio, musicians, engineer, etc.) are paid for by the label. The production fee is therefore left intact as the producer's payment.

CRM Client Record Management system. Commonplace these days in all types of business. The recording industry has a whole host of specialized systems. One of the early versions was Session Tools developed by Apogee with Bob Clearmountain. Current popular systems are headed by Studio Suite (*www.altermedia.com*).

Cue mix Also known as the headphone mix. This is a mix specifically created using pre-fader auxiliaries to provide personalized headphone mixes to musicians.

CV and gate The connections used for analog synthesizers before the introduction of MIDI. The control voltage would provide the pitch while the gate would signal when a note should be produced or not.

DAT Digital Audio Tape. DAT was introduced in the 1980s originally as a consumer format to replace the analog cassette. Protests from the phonographic industry over cloning CDs prevented the format taking off. However, given its potential to record CD quality digital audio, it soon became a 2-track format within the studio.

Data compression The method by which the full data is reduced in a digital file. See formats such as MP3 or AAC for more details.

DAW The abbreviation for Digital Audio Workstation. A DAW is a computer (Mac or PC) usually running some form of audio recording and MIDI editing software such as Digidesign's Pro Tools, Apple's Logic Pro, and Steinberg's Cubase (there are many others).

Depth The space around a recording or mix. In other words, does the music have many layers to it and can some of those parts be heard far away (virtually) while other things are more up front, creating a sound stage as it were? Depth is also referred to as a measure of quality when it comes to the capture of sound. For example, does the instrument have depth and character?

Desk See "Mixing console."

Digital audio A method of capturing an analog waveform by measuring its amplitude against time. The measurement is then held in binary digits and can be then processed within digital signal processing and recorded on computers.

Drum booth See "Booth."

Drums up A description of the order an engineer uses to mix. In this case the engineer will begin with the drums and perhaps work through the rhythm section and end the mix with the vocals.

Dynamic processors Outboard processors (or plugins) which work with the dynamics in signals. Gates, compressors, and limiters are all good examples of dynamic processing.

Dynamics Refers to the *loudness* of the music. If the music is dynamic, it would mean that there is a range of amplitudes from very low to very high. A track lacking dynamics would refer to something that might have been compressed or limited by a dynamic processor.

EBU European Broadcast Union.

Echo An effect where the whole sound is reflected back to the listener after a period of time. Should further delays be added, this is referred to as delay.

Effects processing Effects are applied to signals to provide further character and realism. Reverberation, chorus, and delay are all common effects.

Emotional architecture A term used to describe the emotional profile or shape of a track and includes the peaks and toughs, highs and lows, and moments of tension and release. The emotional architecture of a track is closely linked to various other production elements such as structure, instrumentation, texture, and dynamics.

Engineer The person concerned with the capture and recording of instrumentation and balancing throughout the recording session. The term can be expanded to be an engineer in different parts of the production process such as a mix engineer, mastering engineer, or assistant engineer.

EQ See "Equalization."

Equalization Also known as EQ, equalization is a powerful tool to shape the tonal characteristics of an audio signal. These are commonly found on most audio mixers, both software and hardware varieties.

Expression A term used in music to highlight certain aspects of the performance. In MIDI, expression is a controller that allows you to change the character of the sound to express something within the performance.

Fader A sliding control that governs the amount of level outputted from that channel to the mix (or other selected output).

Far field Refers to sound that is being delivered from a distance. Soffit mounted studio monitors are far field monitors and require more attention to studio acoustics than simply using near field monitors would.

Filter A filter is an audio circuit or plugin algorithm designed to attenuate a certain range of frequencies within the audio spectrum. In its most basic form a filter appears as an LP (low pass) filter where low frequencies are allowed to "pass through" the circuit unaffected while higher frequencies are attenuated or totally blocked from the cutoff frequency. Conversely, an HP (high pass) filter allows high frequencies to pass through the circuit unaffected, while lower frequencies are attenuated or totally blocked. There are other varieties of filter commonly used in both synthesis and EQ applications, such as the band pass and peak/notch filter.

Fix it in the mix An expression commonly used to explain how it is often perceived that a recording can be magically improved in the mix stage alone.

Flange An effect originally created by synchronizing two analog tape machines together with the same audio material playing. By touching the flange (the outer part of the tape reel) and slowing one machine, the audio is temporarily delayed and brought back to speed. This is the origin of flange, and hence its term. Nowadays, we use digital versions to recreate this effect.

Genre A style of music which not only has a musical form, but also a culture. Rock music and folk music are both examples of musical genres.

Get-in The time before the recording session starts where you're able to cart the instruments and amplification into the studio and set up for action.

Get-out The time needed to pack up your gear and move out of the studio complex.

GIFT Get It First Time, or getting the take you want first time you record. For example, always record the guide or scratch vocal as it could simply be the best you'll ever get.

Headphones Often referred to as *cans* within the studio, a studio slang that most likely comes from the fact that they resemble a pair of cans off the side of the wearer's head.

Hearing loss Refers to possible deafness and the issues professionals should be aware of. Exposure to loud sound pressure levels over prolonged periods can induce hearing loss. (See Chapter A-2, Analyzing It.)

High Frequency Also referred to as HF or treble. High Frequency indicates those frequencies above approximately 10 kHz which make up the treble aspects of what we hear.

Holistic listening A way of hearing the whole material at once, with little or no particular focus on any particular element. Being able to switch into this can be beneficial to ensure that an overall and objective picture can be sampled of the mix or recording you're working on.

Inserts These allow the engineer to *insert* a process into the signal flow through a channel strip of a mixing console. One can insert a compressor, for example, into the channel using inserts. Each channel on an analog console has its insert brought out to a patchbay so the signal can be interrupted to insert a gate or compressor, for example.

ISRC International Standard Recording Code. A unique identifier attached to every recording (where applied). Can be placed within the CD subcode. ISRC codes are useful for linking up royalty payments.

ISWC International Standard Musical Work Code is a unique reference number to identify a composition.

Limiter Similar to a compressor although the compression ratio is typically 20:1 and very little signal will go above the set threshold level. Very useful for ensuring levels do not increase over a certain level when recording to digital, perhaps in a live concert recording.

Listening analysis Refers to the art of listening and how we choose to dissect the things we hear. Listening analysis can be a fantastic tool to gauge what has been recorded before and how you may be able learn from others and emulate their styles or tricks within the studio environment.

Live room The space in a studio complex where the musicians are set up to perform. This space is often sound proof and used for recording drums and other loud instruments; it is sonically treated on the inside to ensure an even acoustic response.

Loop The term used to describe a piece of audio that can be repeated over and over in order to make one continuous loop. This can be made up from a huge variety of instruments or parts. Common examples could include a one-bar drum loop or an 8-bar guitar riff.

Lossless A data compression format that does not take away any information from the audio.

Lossy A kind of data compression that deducts data from the original before encoding. When decoded, this information is lost.

Loudness How loud something is in discussion, or a misleading control in hi-fi which increases the bass and treble in the signal. This is in an attempt to overcome the effects of the Fletcher and Munson Equal Loudness contours at lower volumes. Loudness also relates to the perceived loudness in a piece of audio as opposed to its peak value. VU meters are often employed to see how loud a signal actually is.

Low frequency The bass end of material, typically below 120 Hz.

LSD (Lead Singers Disease) A fun, but frequently used, term to describe arrogance, inflated importance, and over-confidence in a lead singer.

Macro listening Listening to the whole track in a certain way to ensure the mix is balanced. Contrast with micro listening (see below).

Meter bridge The block above the top of the channel strips typically containing the VU meters and on which the near field monitors are often balanced.

Micro listening The ability to focus in on certain instruments or frequencies within broadband material (a whole song's mix, for example).

Microphone A transducer that translates acoustic sound into electronic waveforms for use in audio equipment.

Mid field Mid field monitors are those speakers that are not soffit mounted at some distance, but also do not sit in the near field position such as on the meter bridge.

MIDI Musical Instrument Digital Interface, a digital communications protocol between sequencers, synthesizers, and samplers which was incredibly well-designed and is the backbone communication protocol within Reason, Logic, and Pro Tools to speak to its internal synths.

Mix engineer A person who specializes in mixing tracks. The mix engineer is not involved in the recording process and will only have the multitrack sessions sent to them in order to create the final mix (the physical tapes, in analog terms).

Depending on the track and what is required, the mix engineer may or may not have some creative input, such as adding kick or snare samples or creating new textures with chorus, delay, and reverbs, for example. The mix engineer usually comes from a recording engineer background where they have worked closely with various producers.

Mixing console Also known as a board, or desk. The mixing console is the central place for all signals to be blended. It accepts signals from the live room from microphones and can route these to a multitrack and then receive that signal back for mixing.

MP3 Also known as MPEG 1, Layer 3, or MPEG 2, Layer 2. A data compression format working on auditory masking principles to reduce digital audio file sizes. This is a common format used for all portable music players. In recent years MP4 has become more prevalent.

MP4 MPEG 4, layer 14 is an audio format that improves on the MP3 data compression format.

MPC An Akai product (Music Production Centre) that was very common for many programming duties, particularly in R'n'B as well as other genres.

MPG Music Producers Guild.

Multitrack The ability to have more than one parallel track on the tape at the same time. We're now familiar with 24 track 2-inch open reel tape machines such as the Otari MTR90, for example. This has 24 individual tracks on the tape and can allow therefore 24 individual streams of audio recorded at different points. Multitrack has taken on a new form now insomuch as modern systems such as Pro Tools have many more track counts than this.

Music production The art of developing and capturing mainly popular music for broadcast or sale. The onset of recording technology has expanded possibilities, and the music producer develops, directs, and guides the music to the best recorded outcome. This book provides a better explanation in its entirety.

Near field Refers to sound that is produced near to your ears. In most cases this refers to near-field monitors, which are normally placed on a meter bridge in front of you and are considered to be near to your field of hearing. Near field monitors are good in that they reduce the need to acoustically treat the room you're working in.

Networking Networking has two meanings. First it refers to the art of making contacts in public situations and we cover this in detail in Chapter B-1, Being A Producer. Networking can also refer to the physical connections and network made between computers within an environment such as a studio.

NS-10M The Yamaha NS-10Ms have become quite synonymous with many studios. Their white woofers can be seen adorning mixing consoles across the world. These speakers have become popular over the years as they challenge engineers to work hard to make their work sound good. The NS-10s were adopted by the studio community as they reflected the hi-fi quality of the time.

Overdub The process of recording an audio track while other previously recorded tracks are playing back, thus allowing differing parts to be layered over one another without the need for multiple performers, for example the overdubbing of a lead vocal along to a backing track.

P&E Wing Producers and Engineers Wing of the Grammys.

Pan Pan pots, as the controls are often referred to, describe where you'd like a particular channel's output to be placed within the stereo spectrum. The pan will be marked L to R indicating left to right.

Parallel recording Referred to in this book as the choice to record as many instruments at once within the studio, as if it were a live performance by the band. Overdubs can later be added to firm up performances. This is an efficient way of recording if the band are well-rehearsed.

Pass This is the term we use to describe every time the track is played. It refers to the amount of times a piece of tape passes the play head in a tape machine, hence passing the tape head.

Patchbay A set of sockets, usually Bantam within many studios, where all the audio signals in the studio terminate and allow the engineer to patch to pretty much anywhere else. This is a fantastically flexible tool within the studio.

PCM Pulse Code Modulation, the transmission method by which non data-compressed digital audio file formats are communicated such as the CD. PCM is at times used to refer to non data-compressed audio as opposed to MP3 which is data compressed.

Peak The loudest point of a waveform.

Peak meter Illustrates for the user the peak part of the waveform. In the world of digital audio, peak meters are very relevant as they provide a clear indication if you're going to exceed the 0 dB ceiling into digital distortion. However, in earlier times VU meters were more common as analog tape machines could accommodate peaks of in excess of 15 dB, dependent on tape quality.

Perceived production A concept by which the listener can hear a new piece of music and in some way anticipate what the composition will do next. Many rock and pop ballads pull at the same strings. Successful chart-topping material can often follow this style of production.

Phase Often denoted as ø. Phase refers to the polarity of the waveforms. For example, a snare drum can be miked from the top and the bottom. If two microphones face each side of the drum, the top mic's diaphragm will be pulled as the snare skin is hit, while the bottom mic's diaphragm will be pushed in as the snare's bottom skin will push outwards. In real terms this will mean the signal for the bottom snare will be out of phase with the top mic. It is often good practice to record the bottom snare out of phase to rectify this. Phase is also very important to keep an eye out for when working, and correlation meters can provide information about the phase of your material.

Phase effect Where the original signal is merged with a duplicate which is out of phase to the original. This provides a whooshing sound, which is where certain frequencies are in phase and others are not. The whooshing is caused by a low frequency oscillator which dictates the amount of phase and the frequency of the dynamic effect.

Post fader A term used to describe the listener hearing the sound after it has been processed through the channel strip and after the fader.

Postproduction A term used to explain the process that mixes go through after they leave the mix engineer. Also known as *mastering* (see "Mastering"). Postproduction also describes the pulling together of all the elements in a film production.

PPL Phonographic Performance Limited, now referred to as PPL. PPL manages royalties from recorded music in videos, public performance, and broadcast.

Pre chorus As the name suggests, the pre chorus is the section immediately before the chorus, usually following a verse. However, a verse may lead straight to the chorus section without the use of a pre chorus. Generally speaking, the pre chorus is used to set up the chorus section, helping to build tension and heighten the sense of expectation. (A word of warning: the pre chorus is sometimes referred to as the *bridge* in Europe! But do not confuse this with the American use of the term *bridge* which describes the mid section or *mid-8*!)

Pre-delay The amount of time between the sound source appearing and the first reflection from a reverberation. This can be set creatively and in time with your music, or can be manipulated to provide a sense of how large a space is.

Pre-fader A signal that is heard or monitored (such as pre-fader listen) before the fader has had a chance to attenuate the signal. A good example of this is pre-fader auxiliaries where the musicians do not wish to hear the changes the engineer makes in the control room when performing.

Pre-production A term used to describe the planning and preparation, whether in the studio or not, in advance of a recording session. Pre-production is the activity that comes prior to the main recording session. This period is often used to hone the songs and prepare the musicians, studios, and so on for the remaining project.

Presence Refers to whether something is heard up front in a mix, or can be a feature on many pieces of audio equipment such as a guitar amplifier; provides additional presence to the instrument by boosting high mid frequencies.

Printing The term *printing to tape* is often used to describe the rerecording of material that has already been recorded and has been treated. So, for example, committing some guitar effects to tape. Let's print it to tape!

Pro Tools operator A growing profession taking over from the tape operator. A Pro Tools operator is someone who manages the session on the computer. Most engineers will take this duty on unless their attention is completely required on the desk, say in an orchestral session, for example.

Producer agreement An agreement made between the record label and the producer for the production of an artist's record. Essentially the producer is contracted or employed by the label to produce and deliver a set number of tracks from the artist. These can be referred to as *master recordings* or *masters*. (A master usually refers to a song on an album.) A producer agreement will involve either a *costs* deal or *all-in* deal with the label.

Producer's C.A.P. The three pillars of production: Capture, Arrangement and Performance. See Chapter C-4, The Desired Outcome.

Production agreement An agreement entered into by the artist and a production company (usually set up by the producer involved). This acts as a form of recording contract between the artist and production company and producer to make a record. If the project is completed successfully the artist and producer will agree to sell or license the recordings to a third-party label for release.

Production meeting A meeting between the engineering team and producer about the sessions and project. Other people such as A&R person and the musicians might also attend. These may not be formalized and might simply be a chat in the studio complex corridor. A production meeting will usually deal with planning or decision-making.

Project management The intricate planning many businesses engage in when rolling out change or a large project. In music production terms, this is a different flavor, but the skills can be used to assist in the overall success of the project. Many producers will engage in their own variety of this. See Chapter C-3, Project Management.

PRS for Music The name the Performing Rights Society adopted following its merger with MCPS (Mechanical Copyright Protection Society). Mechanical royalties refer to the sales of an artist's (or composer's contribution to) CDs but also when their music is used for television, film, or an advert. Performance rights refer to royalties paid to the artist (or composer) of a piece of music each time it is played.

Pulse Code Modulation See "PCM."

Record deal A contract between a record label (usually) and the artist that stipulates what the artist will produce for the label and what the label will provide financially in an advance to make this possible. See Chapter C-1, What's The Deal.

Recouping A term used to describe the process of regaining money spent by a record label on recording and other associated costs. These sums are usually taken out of the artist's royalty share and therefore the artist will not start to receive any income from their work until they have recouped all the costs for their label.

Reference mix A rough mix produced by the recording engineer or producer and sent to the mix engineer in order to give a direction or flavor as to where the track needs to go sonically. It is intended as a guide to the final mixing process.

Reverb time Also known as the reverberation time, it dictates the time it takes for the reverberation to reduce by 60 dB. This indicates to the listener how big the reverberation is. It might signal that the room is large, in conjunction with the pre-delay.

Reverberation A natural phenomenon of sonic reflections from surfaces in a room or elsewhere. These characterize the sound and the environment they're placed in. Modern digital reverberation units allow us to electronically place any instrument in any space and tailor it to the music we're working on.

Rhythm section The group of instruments that typically contains the drums, bass, and percussion, plus any other rhythmical backing instrumentation. This describes the backline in the recording, or the backing band if you like.

Royalties Refers to the income generated by the various streams of revenue for the record label, artist, songwriters, producer, and musicians.

RT60 Often referred to as T 60 (see "Reverb time").

Serial recording Refers to the choice to record each instrument one by one, or similarly as opposed to all at once. This is common practice when writing in the studio, or building up music for a singer songwriter without a band, for example.

Session management Refers to a number of points, namely the management of the session and the personnel involved in aspects such as timings, personnel, equipment, and so on. Session management also encompasses things such as taking notes, taking responsibility for the data you'll need later, such as who contributed to which part, who performed what where, and so on.

Session musician A musician hired to perform on a record or performance who may not be part of the band or be associated with the act. Session musicians are usually very highly adept on their instrument and can be extremely versatile and creative within the studio.

Session notes Any notes you feel pertinent to take down during a session. There will be decisions you make which you'd like to keep for later on during the recording process. Or you may wish to note down why things have happened the way they have. Anything you need to take down and perhaps share.

Sight reading Classical musicians spend many years training to a very high level to be able to read musical scores and to perform them live. This ability, while not essential for the producer of popular music, is essential for that of a classical music producer. Session players usually have highly tuned sight-reading skills and some more classical players may insist on a score before playing.

Solo in place A feature on many mixers where when you solo a particular channel, say an acoustic guitar, the effects associated with that track remain in place so you can hear the amount of reverb on that instrument.

Sound Pressure Level Also referred to as SPL, Sound Pressure Level is how loud something is in the air as opposed to within audio equipment.

Spill A term used to express the sound of other instruments captured by another microphone. For example, the acoustic guitar microphone might pick up spill from the piano within the same room.

Studio complex The suite of recording studios and rooms that make up the whole studio facility, which might contain four recording studios, a mastering suite, and some programming rooms.

Studio time The term used to refer to a studio booking, or the amount of time scheduled to take place in the studio session.

Sub Woofer A speaker employed to deliver just the very lowest frequencies going down to 20 Hz in a monitor system. Systems that denote 2.1 or 5.1 indicate they have a subwoofer by the ".1".

Take A capture of a performance. We speak of take 6 being the best. This would be the 6th recording of a particular track (on tape). Engineers document takes on take sheets, although modern recording devices such as Pro Tools and Logic, for example, do not require this so much, as the notes can be made on screen.

Take sheet A document used to list the takes and when they start.

Tape operator (tape op) A Tape Operator's responsibility is to manage the analog or digital tape through a recording session. Their role is to choose which track to record on and operate the machine's transport controls. This role has been phased out over the years and replaced by simply *assistant engineer* or *Pro Tools operator*.

The Big Four Refers to the big four record labels (Sony Music Entertainment, Universal Music Group, EMI Group, and Warner Music Group).

Tie line The connection made across a building. For example, there may be tie lines in a studio complex that link Studio 1's live room to Studio 3's control room for maximum flexibility.

Total recall The ability to save the various settings and positions of all the controls on a mixing console, thus saving the huge amount of time that was once required to set up or reset a desk before or during a session. SSL was one of the first desk manufacturers to introduce a total recall function in the 1980s.

Track The song or each track of a multitrack tape, or channel on a console.

Track sheet This was useful to tell engineers what tracks were recorded on each open reel tape (or digital tape, DTRS). In the days of the analog tape, the back of the box often provided the best space for this.

Transport/transport controls The term used to describe the movement of tape, whether analog or digital. Tape transport controls, for example, refer to the play, stop, record, fast forward, and rewind controls. The transport mechanism refers to the motors, spindles, and so on that drive the tape through the machine. In modern times, transport controls are the same, but no longer refer to the movement of tape.

Vocals down A description of the order an engineer starts to mix. In this case the engineer will begin with the vocals and perhaps work down, ending with the rhythm section.

VPL Video Performance Limited, set up to collect royalties for music videos played on TV stations such as MTV.

VU Volume Unit, a measurement of loudness used on many mixing consoles and audio equipment. Peak meters are more common these days, but increasingly VU is becoming an ally as it illustrates the average loudness of a signal rather than just its peak.

Wall of sound The watermark of producer Phil Spector, who created large and lush performances from a wide range of musicians using careful arrangements and placements of musicians within the live room.

WAV The .wav format is a non data-compressed digital audio file format used on computers.

Width Often refers to the stereo field and the width across this.

XLR The connector on most professional microphone cables. XLR plugs come in female and male varieties and are commonplace for microphone cables and AES/EBU digital audio connections.

Index

Note: Page number followed by *f* indicate figures; *b* indicate boxes.